8th World Congress on Particle Technology 2018

AA009144

Keynotes and Plenaries

Orlando, Florida, USA
22 - 26 April 2018

ISBN: 978-1-5108-6976-9

Printed from e-media with permission by:

Curran Associates, Inc.
57 Morehouse Lane
Red Hook, NY 12571

Some format issues inherent in the e-media version may also appear in this print version.

Copyright© (2018) by AIChE
All rights reserved.

Printed by Curran Associates, Inc. (2018)

For permission requests, please contact AIChE
at the address below.

AIChE
120 Wall Street, FL 23
New York, NY 10005-4020

Phone: (800) 242-4363
Fax: (203) 775-5177

www.aiche.org

Additional copies of this publication are available from:

Curran Associates, Inc.
57 Morehouse Lane
Red Hook, NY 12571 USA
Phone: 845-758-0400
Fax: 845-758-2633
Email: curran@proceedings.com
Web: www.proceedings.com

TABLE OF CONTENTS

Some abstracts in this book refer to images that were not made available by the publisher

508388 An Empirical Comparison of Two Different Cyclone Designs in the Usage of a Third Stage Separator

Abstract:

Background and Motivation

Actually the Third Stage Separator (TSS) in a FCC-process is mainly constructed as a standard reverse flow cyclone which can be designed by approved calculation models, as for example by using the model of *Barth/Muschelknautz* [1]. The FCC-process efficiency mainly depends on the cyclone performance i.e. on its separation efficiency and its pressure loss. The conical shape of the lower separation body of a reverse flow cyclone leads to high tangential velocities and therefore to high erosion which strongly affects its life time. A second cyclone type, which is not yet well investigated, is the uniflow cyclone. Its essential advantages compared to the reverse flow cyclone are its compact construction and its easy integration into pipes. A further advantage is its low pressure loss and therefore its low energy consumption. *Postma et al.* investigated the usage of uniflow cyclones (swirl tubes) in the FCC-Process, which requires high process reliability [2]. A study on the applicability of uniflow cyclones as a TSS was carried out in 2010 at PSRI laboratories and showed acceptable performance data for the usage within FCC-process engineering [3].

Standard Reverse Flow Cyclones vs. Uniflow Cyclones

The simplicity of design and the lack of moving parts in cyclone separators are key to their ubiquitous presence in process industry. In some applications, where high gas flow rates must be processed, the physical size of a conventional reverse flow cyclonic unit can become very large. The uniflow cyclone or axial cyclone, is a interesting alternate design, which is able to deliver comparable efficiencies in a very compact geometry. The name is derived from the flow field pattern generated inside the cyclone. Unlike reverse flow cyclone, the gas enters and leaves the cyclone in the same direction (see Figure 1).

Fig. 1: Depiction of a uniflow cyclone and its flow field pattern.

Compared to a reverse flow cyclone design, the uniflow cyclones can be built approximately two-third smaller in volume at comparable efficiencies. Wide spread adoption of this design has been limited by the absence of a universally accepted and validated design approach. The existing literature mainly focuses on special applications, as shown in the research work of *Gauthier et al.* or *Zhang et al.* [4,5]. The highly turbulent particle flow patterns offer a complex parameter system which makes the dimensional analysis rather complicated. For instance, numerous design variations of the inlet swirl generator (Figure 1) are possible, and each approach will result in remarkably different flow field within the cyclone. Therefore, an empirical approach is most apt for studying uniflow cyclones.

To facilitate the understanding of its operation, the design structure of a uniflow cyclone can be divided into four major sections in the direction of the flow, namely Swirl Vane Inlet, Separation Chamber, Vortex Finder and Particle Outlet.

State of the art

Even though the working principle of this cyclone type is well-known, there is hardly any literature which investigates the design, the estimated efficiency or pressure drop. To close this gap the MCI is driving systematic research activities since 2008 to improve the separation efficiency of uniflow cyclones by optimizing the geometric design [6,7,8,9]. As outcome of these activities, there are several applications established in process and automotive industry, as for example in air-pre-cleaning systems for internal combustion engines used in off-road vehicles (e.g. building and agricultural engines).

Experimental

To compare the two cyclone systems on a fair basis, the executed test series show data for efficiency and pressure drop and the erosion potential at comparable operating conditions. The experimental testing is performed on a standard reverse flow cyclone and a uniflow cyclone with various operation configurations differing from each other regarding the vortex finder diameter, the volume flow rate and pressure recovering systems. A test stand which allows comparing effectiveness and optical accessibility for phenomenological observations is set up at MCI laboratories, by using selected FCC-Powder which typically reaches the Third Stage Separator.

References:

[1] *VDI Heat Atlas,* 11th Edition. VDI-Book. Berlin [u.a.]: Springer Vieweg, 2013.

[2] Postma, R.S., Hofmann, A.C., Dries, H.W.A., Williams, C.P., The Use of Swirl Tubes for Dedusting, World Congress on Particle Technology 3, Brighton, UK, 1998.

[3] Kraxner, M., Muschelknautz, U., Karri, S.B.R., Cocco, R., Knowlton, T.M., Applicability of a Uniflow Cyclone as a Third Stage Separator in the FCC-Process, AIChE - American Institute of Chemical Engineers - Annual Meeting, Minneapolis / MN, USA, 2011.

[4] Gauthier, T.A., Briens, C.I., Bergougnou, .A., Galtier, P., Uniflow cyclone efficiency study, Powder Technology, Nr.62: 217-225, 1990.

[5] Zhang, Y., Wang, X., Riskowski, G.L., Christianson, L.L., Ford, S.E. Particle separation efficiency of a uniflow deduster with different types of dust, Transactions of ASHRAE, 1999.

[6] Muschelknautz, U., Pattis, P., Reinalter, M., Kraxner, M., Design Criteria of Uniflow Cyclones for the Separation of Solid Particles from Gases, CFB10 â€“ 10th International Conference on Circulating Fluidized Beds and Fluidized Bed Technology, Sunriver / OR, USA, 2011.

[7] Kraxner, M., U. Muschelknautz, S. Wechner, S. Ackermann, V. Greif, J. Bolda, Influence of the Inlet Vane Geometry on the Uniflow Cyclones Performance, AIChE-Annual Meeting, Pittsburgh / PA, USA, 2012.

[8] Kraxner, M., Skarke, B., Kofler, T., Pillei, M., Pressure Drop in Uniflow Cyclones: Investigation on an Empirical Calculation Model, CFB-11, Beijing, China, 2014.

[9] Kraxner, M., Empirical Evaluation of Design Criteriaâ€™s for Uniflow Cyclones in Multicycloneboxes, Doctor Thesis, TU MÃ¼nchen, Germany, 2013.

Copyright © American Institute of Chemical Engineers. All rights reserved.

506435 Numerical Study of Hygroscopic Particle Removal Using a Cyclone Separator with Saturated Water Vapor

Abstract:

Introduction: Cyclone separator is one of the most widely applied equipments for separating particles from the air. However, it is difficult to separate particles with diameters smaller than 2.5 micron using the cyclone separator. In light of the particle-vapor interaction, it is possible to increase the particle diameter via forced condensation, especially when the particle contains soluble component. This study has examined the possibility of using a cyclone separator to collect the soluble particle, i.e., sodium chloride (NaCl) as test example, numerically. Specifically, a steam of saturated water vapor with elevated pressure (0.7 MPa) is introduced vertical to the air-particle inlet to increase the particle size by hygroscopic growth, i.e., forming a NaCl-water droplet due to absorbing water vapor. Comparisons of trajectories between the particles with and without hygroscopic growth are presented.

Methods: A cyclone separator model and its mesh, i.e., tetrahedral mesh with prism layers, were created (see Fig 1.). The velocities of the air and the water vapor were 10 m/s and 1 m/s, respectively. The NaCl particle density was 2165 kg/m^3. The Euler-Lagrange method was applied to predict the airflow and particle transport. If The NaCl-water droplet collided with the cyclone surface, deposition was assumed to occur when considering the particle-vapor interaction. For comparison, the particle separation without hygroscopic growth was also simulated to examine the separation efficiency. Simulations were carried out by ANSYS Fluent with user-defined functions based on validated particle-vapor interaction model (J. Aerosol Sci. 2017, 105: 108-127).

Results and Discussion: The trajectories of the 2.5 micron NaCl particles are shown in Fig. 2 when including the effect of particle-vapor interaction. The particle / droplet size increases rapidly when traveling through the air with high relative humidity. Thus, these large droplets would deposit on the cyclone surface due to inertial impaction, which leads to a high separation efficiency. The solution could be collected at the bottom of the cyclone separator with extra water trickled from the top of the cyclone, if the deposited droplets and condensed water cannot form a liquid film on the cyclone surface. Fig. 3 illustrates the trajectories of the 2.5 micron NaCl particles without hygroscopic growth.

Fig.1 Geometry and mesh for the cyclone separator

Fig. 2 2.5 micron particle trajectories with hygroscopic growth

Fig. 3 2.5 micron particle trajectories without hygroscopic growth

Conclusions: The numerical study validated the feasibility of separating small soluble particle (<2.5 micron) using cyclone separator by introducing the saturated water vapor with elevated pressure. This method could be useful for the design of the fine particle removal device with simple structure and easy-maintenance.

Acknowledgements: The authors gratefully acknowledge the financial support of the National Natural Science Foundation of China (grant No. 51606041), National Natural Science Foundation of Jiangsu Prov. (grant No. BK20160688). The Fundamental Research Funds for the Central Universities are also acknowledged.

Copyright © American Institute of Chemical Engineers. All rights reserved.

DESIGN CRITERIA FOR MULTI CYCLONES IN A LIMITED SPACE

Ulrich Muschelknautz, MK Engineering, Heidelberg, Germany

Abstract

Multi cyclones for gas-solids separation consist of many parallel cyclone cells within a common housing having a common solids hopper. They are preferably used for high-efficiency separation of a gas-solids flow. Principally increasing the number of cyclone cells and decreasing their size improves the efficiency of a multi cyclone without changing its pressure drop, provided a uniform distribution of gas and solids feed into each single cyclone cell can be achieved and bypass flows through the solids discharge openings from one cyclone cell to the other can be avoided.

In many cases multi cyclones use standard reverse flow cyclone cells with an axial inlet, also called swirl tubes. Analogously multi cyclones can be made from uniflow cyclone cells. Deviating from swirl tubes uniflow cyclones have gas and particles passing through them in only one direction exiting at the same end of the device which makes them much more compact than swirl tubes. In recent years comprehensive experimental and theoretical studies of uniflow cyclones have strongly improved their understanding and led to approved design criteria and calculation methods similar to those which have already been achieved for standard reverse flow cyclones. In the present investigation well proven calculation programs have been applied to compare multiclones based in swirl tube cells with those based on uniflow cyclone cells regarding their efficiency to separate a specified gas-solids feed at a given pressure drop. The investigation has been performed by varying the available space for the dedusting device, the number of cyclone cells, the cyclone pressure drop, the particle size distribution of the feed and the particle density. The results indicate that, under the considered operation conditions, a uniflow multiclone is more efficient than its swirl tube counterpart when the space is limited onto about the volume needed by the uniflow multiclone for its optimum performance and the available pressure drop is low.

KEYWORDS

Multi cyclone, multiclone, uniflow cyclone, standard cyclone, swirl tube, axial cyclone.

Introduction

Cyclone separators are widely used in industry for separating solid particles from gases. They are applied preferably for reasons of high operational safety due to their simple and robust construction, usability for product recovery and in many applications comparable low invest and operation costs. In addition, they can be operated under extreme conditions including large gas volume flow rates as well as high solids concentrations, temperatures and pressures, representing a comparative advantage over other gas-solids separation technologies. The minimum particle size that can be collected by a cyclone generally decreases with decreasing cyclone size. This effect is used in multi cyclones, i.e. systems of many parallel cyclone cells within a common housing having a common solids hopper for solids discharge. Multi cyclones are generally used to increase the separation efficiency beyond the level which is achievable by a single cyclone. Principally increasing the number of parallel cyclone cells and decreasing at the same time their size improves the efficiency of a multi cyclone without affecting its pressure drop, provided a uniform distribution of the gas and the solids feed into each single cyclone cell can be achieved and bypass flows through the solids discharge openings from one cyclone cell to the other can be avoided.

In many cases multi cyclones use standard reverse flow cyclone cells with an axial inlet, also called swirl tubes, which are shown in Fig. 1 left (In the strict sense of the word swirl tubes denote that type of a cyclone with a pure cylindrical shape (1)). Analogously multi cyclones can be made from uniflow cyclone cells, Fig. 1 right. Often those devices are used in space limited applications, e.g. as third stage separators in fluid catalytic cracking (FCC) processes (2). In both cases axial inlets with swirl vane inserts for vortex generation are used. The main difference between both cyclone types is that in swirl tubes the gas flow reverses its direction and pure gas and collected particles are leaving the device at opposite sides, whereas uniflow cyclones have gas and particles passing through them in only one direction exiting at the same end of the device. Due to those design differences unifow cyclones can be built much more compact and can easily be integrated into pipe lines.

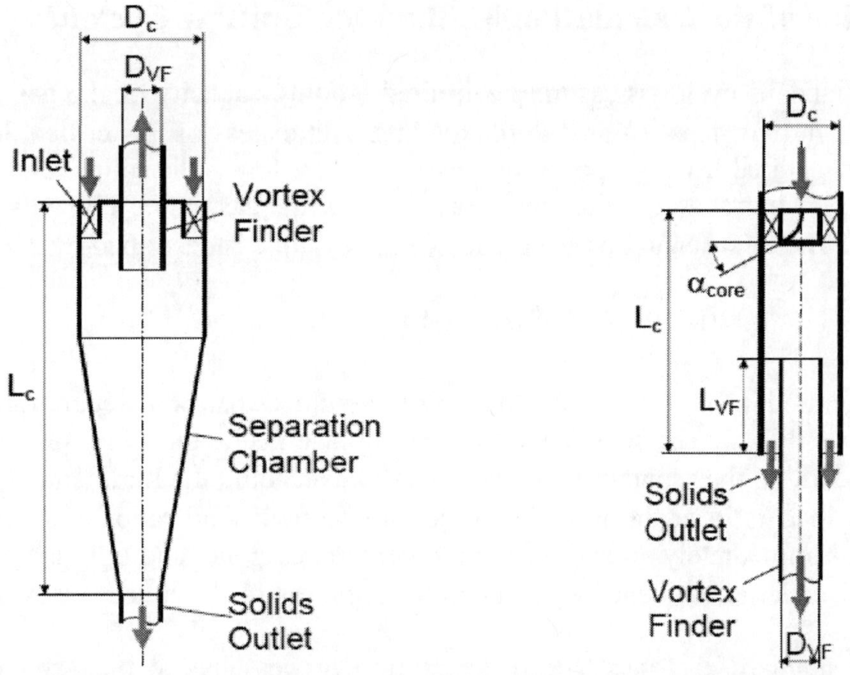

Fig. 1. Schematic illustration of a reverse flow cyclone with axial inlet (left), often applied in multi cyclones, and a uniflow cyclone with axial inlet and vertical downward orientation.

Even though the principle of uniflow cyclones has been known for a long time, not much literature has been published on how to design and to calculate them. So far the investigation of uniflow cyclones has been restricted to special applications such as a short-contact time reactor for high solids loadings (3). To fill this gap systematic experimental studies have been performed recently (4-16). A major challenge to make uniflow cyclones accessible to general industrial practice is an accurate, reliable and fast design method. For calculating the performance data of conventional reverse flow cyclones, approved analytical models with high precision exist which are often preferred over methods based on numerical simulation with long computing times. With this background, a new analytical approach for calculating the separation efficiency and the pressure drop of uniflow cyclones for practical design work has been developed (17, 18). This approach is based on an equilibrium orbit concept similar to that applied in the Barth-Muschelknautz model for reverse flow cyclones, which has been proven to work successfully in a wide range of industrial applications. More detailed information about this model can be found e.g. in (1, 19, 20).

Applying the calculation models for standard cyclones and for uniflow cyclones which both are based on the same physical concepts indicated that uniflow cyclones need much less space for separating a specified solids gas mixture than standard reverse flow cyclones but are generally also slightly less efficient, at least in their present state design (21,22).

Often space for a dedusting unit in an industrial plant is limited. If saving of space is of high priority or if there is even only a limited space available for purifying a given gas volume flow the question arises whether under that restriction of *limited space* a multicyclone made from uniflow cyclone cells may be more favorable than its swirl tube counterpart. This question is addressed by applying well proven calculation programs for standard cyclones according to (19,20) and for uniflow cyclones as described in (17,18).

Outline of the Calculation Method for Uniflow Cyclones

The model for standard cyclones assumes a limited loading capacity of the gas stream splitting the separation procedure in two steps. At any solids loading μ in excess of this critical loading $\mu_{\lim}$, the solids are immediately separated from the gas at the inlet to the cyclone. The solids remaining in the gas are separated in the cyclone barrel and in the inner vortex below the gas outlet tube with a second, inner separation efficiency η_i. The total separation efficiency of the cyclone is described by the equation

$$\eta = (1 - \mu_{\lim}/\mu) + (\mu_{\lim}/\mu)\eta_i \tag{1}$$

Analogous to the model for standard cyclones, it is assumed that both separation processes also occur in uniflow cyclones, Fig. 3. The first separation takes place inside the swirl vane inserts for swirl generation and subsequently in the separation chamber due to exceeding the limited load ratio $\mu_{\lim}$ of the uniflow cyclone. If the load ratio at the inlet μ exceeds the limited load ratio, $\mu_{\lim}$, the excess mass fraction will be removed immediately after the gas jet enters the cyclone, and only a small fraction that is restricted by $\mu_{\lim}$ will undergo the centrifugal separation process in the inner vortex of the cyclones, Fig. 3.

Deviating from standard cyclones a third separation process has to be taken into account in uniflow cyclones: In contrast to standard cyclones the solids discharge in uniflow cyclones is located closely to the gas exit. Furthermore, a considerable part of the gas flow passes through the ring chamber between the cyclone wall and the vortex finder pipe before it leaves the device through the gas exit pipe. Due to those characteristics particles carried into the ring chamber can be re-entrained back into the gas outlet. The re-entrainment rate depends on the cyclone geometry and on the operation data. The efficiency of the gas particle separation within the ring chamber is called η_{RC}. Finally, the collection efficiency of the bunker η_b has to be taken into account, as in standard cyclones. Thus, the total separation efficiency as well as the fractional separation efficiency of the cyclone are functions of those four single separation efficiencies.

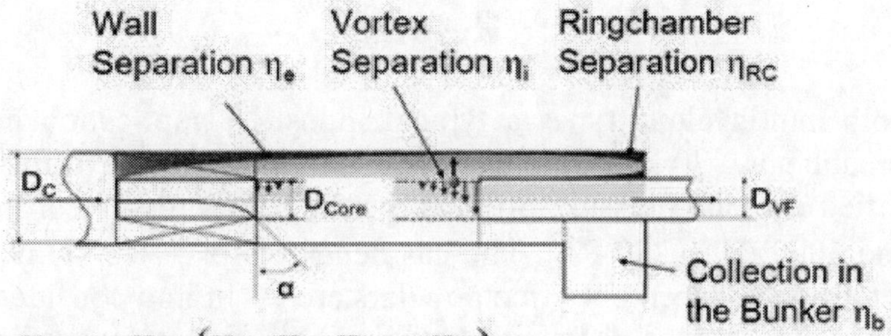

$$\eta_{tot} = \eta_{tot}(\eta_e, \eta_i, \eta_{RC}, \eta_b)$$

$$\eta_{F,tot}(d) = \eta_{F,tot}(d, \eta_e, \eta_i, \eta_{RC}, \eta_b)$$

Fig. 2. Overall structure of the separation process in a uniflow cyclone.

The calculation of the *separation in the inner vortex* η_i has been described in detail elsewhere (17,18). For low loadings μ close to 0.001 this separation mechanism is dominant and its efficiency is expected to be close to the total separation efficiency. With increasing solids loadings the wall separation becomes more and more important and already below a solids loading of 0.01 the wall separation can become the dominant separation mechanism depending mainly on the cyclone size, the tangential velocity at the inlet and on the size distribution and density of the particles.

In uniflow cyclones with an axial inlet the gas flow is set in rotation within a very short time. The entering particles, driven by the drag force of the gas flow, follow this rotational motion, but with a certain delay which is determined by the particle relaxation time. At the end of the separation chamber it may occur that some particles will still not have the circumferential velocity of the gas. This applies in particular to coarser particles with a relaxation time similar to the residence time of the gas within the device. Thus, for small uniflow cyclones and not too fine particles that slip between the circumferential gas velocity and the particle velocity within the separation chamber can be significant and has to be taken into account when calculating the separation efficiency.

Analogous to the calculation model for reverse flow cyclones the pressure drop of uniflow cyclones is calculated as the difference of the total pressures between a position in front of the cyclone inlet and a position m far beyond the opening of the gas outlet where the swirl strength of the vortex flow in the gas outlet has decreased to approximately zero due to wall friction. As in standard reverse flow cyclones the total pressure drop is divided into three parts, the pressure drop in the inlet, the pressure drop in the separation chamber between the mean entrance radius and the position i at the vortex finder radius r_{VF}, and the pressure drop between the position i and the measurement position m in the gas outlet tube. A detailed description of the three pressure drop parts is given in (18).

In the present study the performance data of the considered multi cyclones are determined by calculating the performance data of single cyclone cells. This means a uniform distribution of the gas solids feed onto each cyclone cell has been assumed and bypass flows between the solids discharge openings of neighboring cyclone cells have been excluded.

Results

To compare both multicyclone types a typical industrial application has been considered and investigated in a systematic way: 30 kg/h fine particles shall be removed from 14,000 m^3/h exhaust gas of a reactor at a specified cyclone pressure drop. Regarding the pressure drop two cases are studied, 1000 Pa (0.145 Psi) and 3000 Pa (0.290 Psi). The gas density is $\rho_g = 1.2$ kg/m^3 and the gas viscosity is $2 \cdot 10^{-5}$ Pa s. Regarding the particle feed only fine powders are taken into consideration in order to make a meaningful comparison. Note, that the difference between both cyclone types becomes most clear when considering the collection of fine particles. With increasing particle size of the feed the efficiencies of both cyclone types increase and converge (20). Four fine powders are considered characterized by two different particle densities ρ_s = 1490 kg/m^3 and ρ_s = 2700 kg/m^3 and two different particle size distributions $Q_3(d)$ which are assumed to be describable by RRSB-functions. The RRSB-parameters are (1) d_{min} = 0.1 µm, d_{max} = 30 µm, d' = 19 µm, n = 0.8, d_{50} = 12.2 µm and (2) d_{min} = 0.2 µm, d_{max} = 200 µm, d' = 25.5 µm, n = 0.95, d_{50} = 17.5 µm.

A multi cyclone variant often applied to solve such a problem is Variant ST16-L shown in Fig. 3, left. Its elementary cyclone cells are identical swirl tubes operated in parallel and properly designed for the given gas volume flow. The swirl tube design with an axial inlet has the advantage (compared to the standard cyclone design with an tangential inlet) that those cells a) are more compact (diameter and height) and b) can be packed very densely with very narrow distances between neighboring cells. The characteristic of a swirl tube multi cyclone is that the pure gas and the collected solids leave the device at opposite sides. The pure gas pipelines cross the crude gas inlet duct before they are led into a common clean gas discharge channel located on the roof of the device. The collected solids fall into the common hopper from where they are discharged e.g. by a rotary valve. The walls of the hopper have to be inclined steep enough to prevent caking of particles on the walls. This condition requires a certain height of the hopper and creates thus a large volume of the construction used only for discharging collected particles with no other function else.

A more compact variant for cleaning the same gas-solids feed at the same pressure drop is shown in Fig. 4. This variant, made of uniflow cyclone cells, has both, gas and collected particles, exiting at the bottom of the device, saving considerably height. In addition, due to the small diameter of the uniflow cyclone cells they need less base area than the swirl tube based variant ST16-L. In this construction the pure gas pipelines cross the particle hopper before entering a common pure gas discharge plenum. Caking of particles at the pipelines within the hopper can be avoided by appropriate claddings. The distance between neighboring uniflow cyclone cells has been chosen larger than in the case of the swirl tube multiclone (see Fig. 3, left and right) in order to minimize bypass flows from one uniflow cyclone cell into the other.

The calculation results indicate that both multicyclone variants (ST16-L and UC16) have similar efficiencies, if the allowed pressure drop is 1000 Pa (see Fig. 5). However the uniflow variant needs considerably less space and construction material than the swirl tube cyclone variant, see Fig. 3 and Fig. 4. To see whether a smaller swirl tube variant might compensate that disadvantage without losing too much efficiency swirl tube variant ST16-S with the same base area as the uniflow cyclone variant has been investigated in addition (Fig. 3, center). Note, that for this purpose the smallest possible diameter of the swirl tube cells has been chosen which just allows for complying with the required low pressure drop of 1000 Pa. This variant, however, provides significantly lower efficiencies than the uniflow cyclone device, see Fig. 5.

If a higher pressure drop of 3000 Pa is allowed the results shift in favor of the swirl tube multiclone: In that case its efficiency is in all considered cases (varied particle sizes and particle densities) higher than that of the uniflow multiclone, see Fig. 6. This is true also for the small swirl tube variant ST16-S having the same base area as the uniflow multiclone.

Increasing the number of cyclone cells from 16 to 30 (for purifying the same gas-solids feed) and performing an analogous investigation for those more complex variants indicate similar results as for the 16-cell variants, but with all efficiencies increased compared due to the higher cell number (Figs. 7-10).

There seems to be no principal difference between a multiclone based on swirl tube cells and a multiclone based on uniflow cyclone cells when it comes to erosion in the entrance zone. Both cyclone types have the same concept of swirl generation which is realized by swirl vane inserts. Critical erosion zones are in both cases the blades of the swirl vane inserts as well as the cyclone wall in the entrance region. Due to the smaller cyclone diameter erosion in the uniflow cyclone cells may be more pronounced than in their swirl tube counterpart if the entrance velocity is in both devices the same. However, slightly reducing the entrance velocity in the uniflow cyclones can make that difference disappear.

A uniform distribution of the gas and particles on all cyclone cells is in both multiclone types challenging and can principally be solved with the same concepts in both cases, e.g. by declining the separation wall between the common upstream and downstream plenums as is shown in Fig. 3.

The pressure drop of both multiclone types can be reduced very efficiently and by about the same amount by installing additional swirl vane inserts into the gas outlet pipes of each cyclone cell (23). Those curved blades transform the rotational flow entering the gas outlet pipe into a pure axial flow and recover in this way rotational energy of the gas flow back into static pressure. Without this measure that energy is lost due to dissipation. This measure has been successfully applied in numerous industrial cyclones in a wide range of industrial applications, see e.g. (24).

Multiclones with 16 Cells

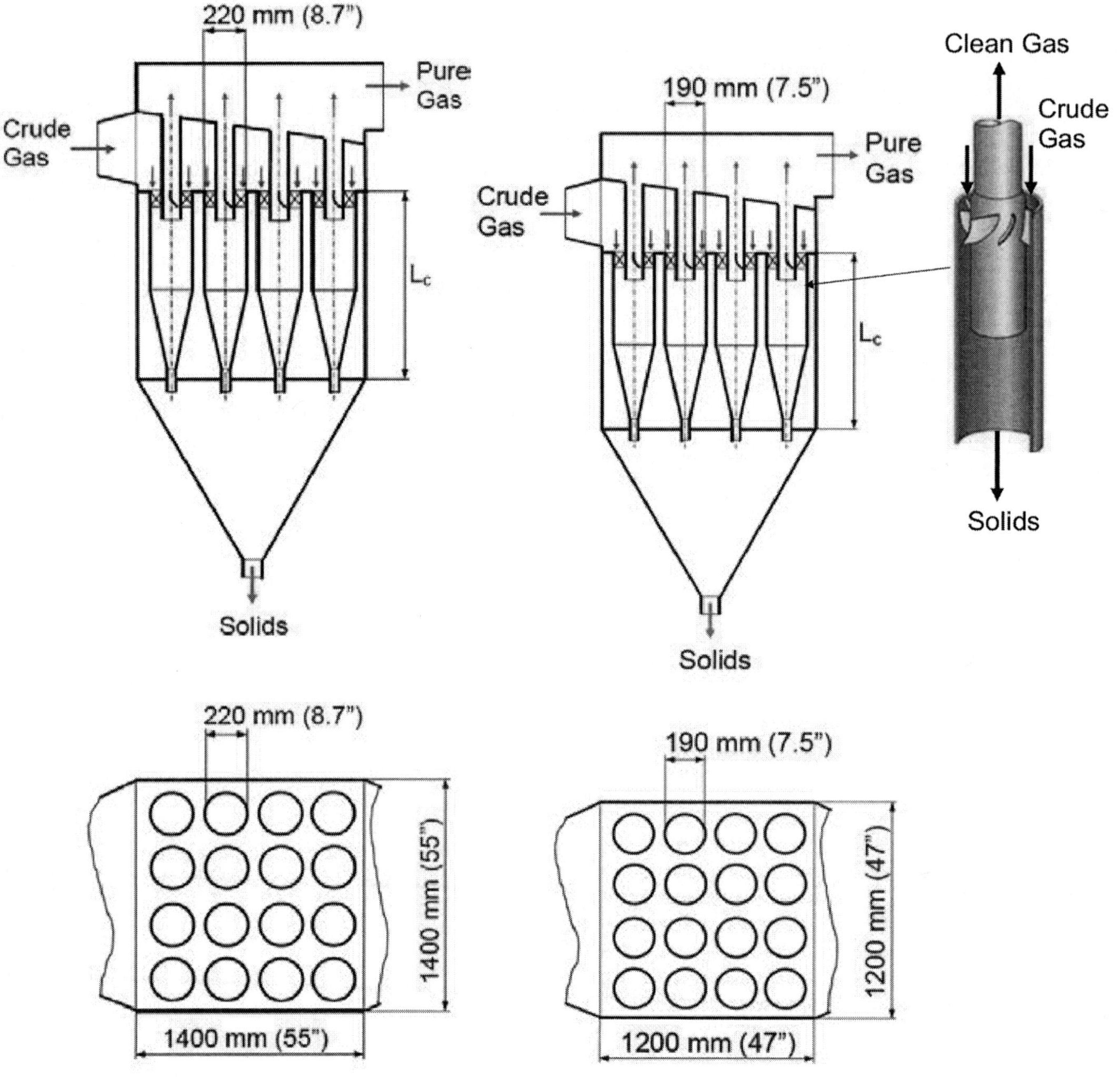

Fig. 3. Multicyclone-Variants ST16-L (left) and ST16-S (right) with 16 Swirl Tubes.

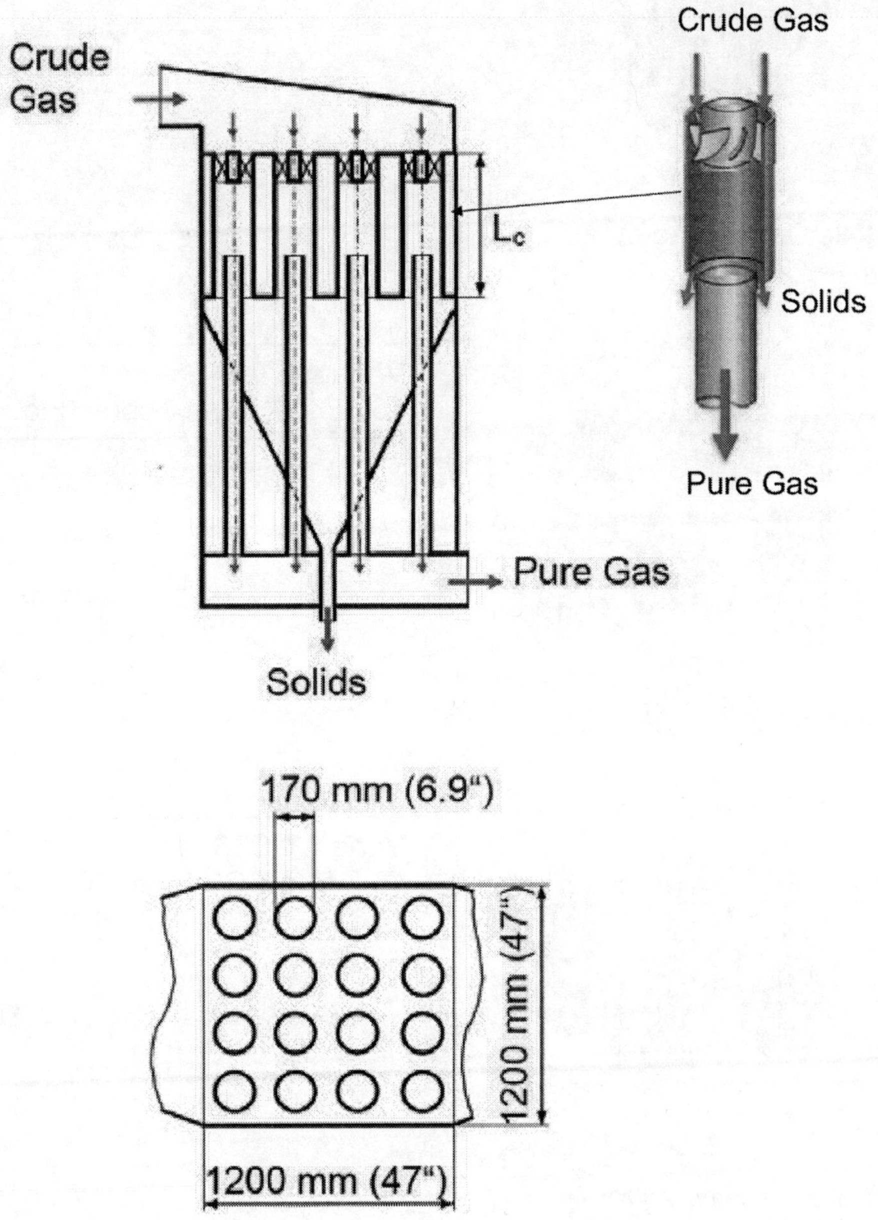

Fig. 4. Multicyclone-Variant UC16 with 16 Uniflow Cyclones.

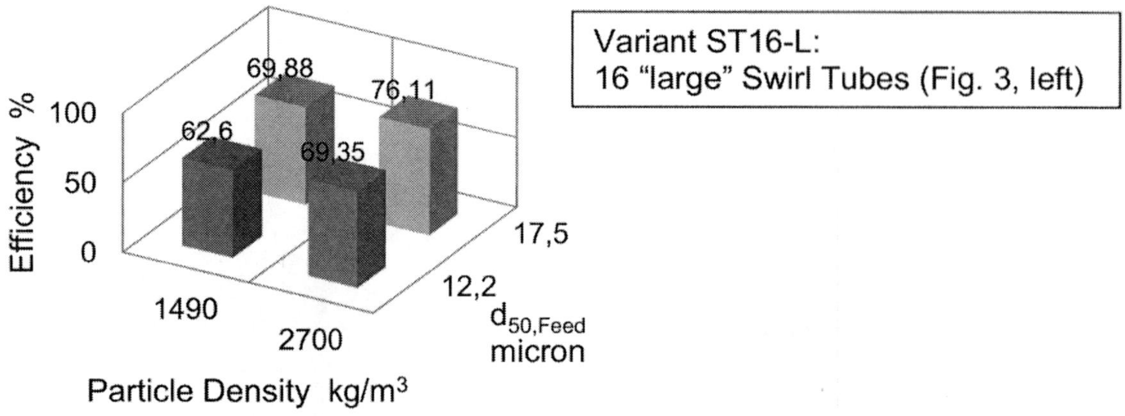

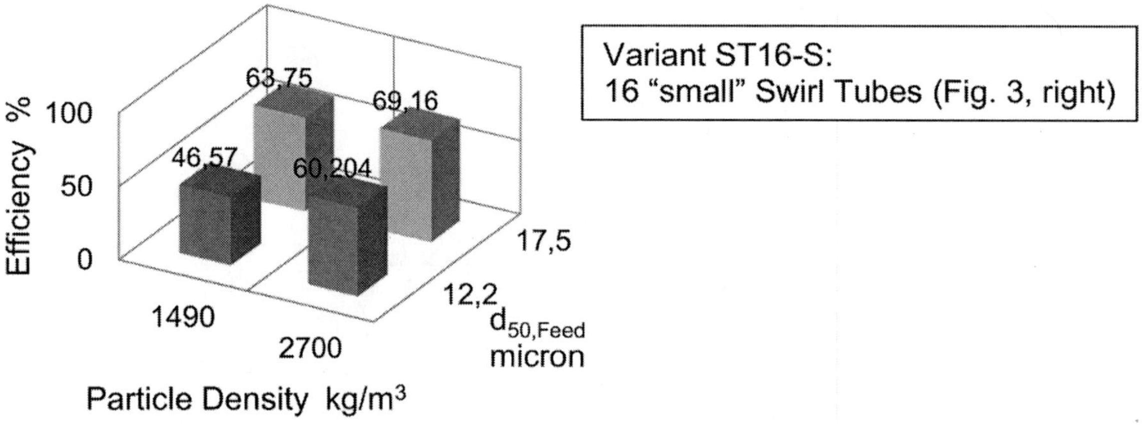

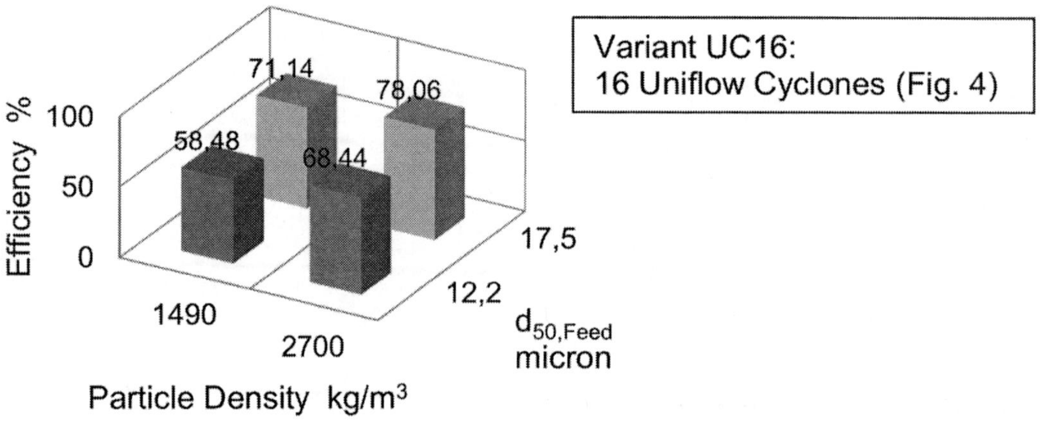

Fig. 5. Calculated Separation Efficiencies of Multicyclone-Variants ST16-L (top), ST16-S (center) and UC16 (bottom) with $\Delta p = 1000$ Pa.

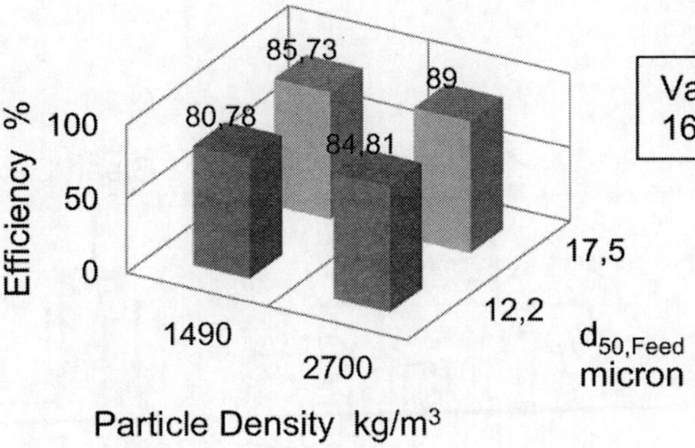

Variant ST16-L:
16 “large” Swirl Tubes (Fig. 3, left)

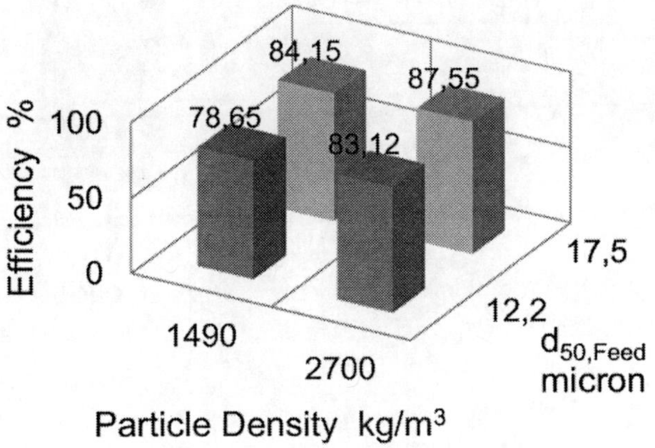

Variant ST16-S:
16 “small” Swirl Tubes (Fig. 3, right)

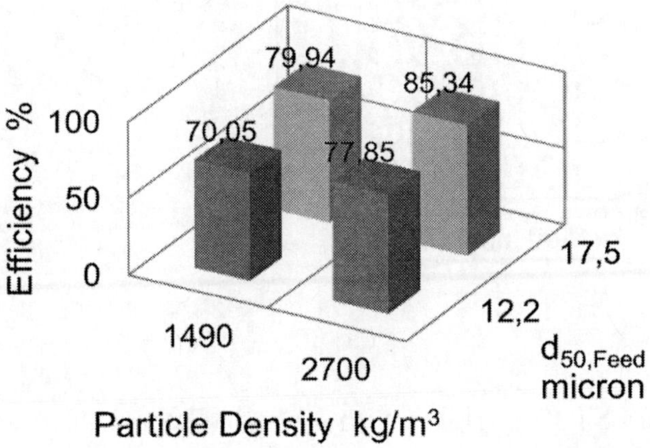

Variant UC16:
16 Uniflow Cyclones (Fig. 4)

Fig. 6. Calculated Separation Efficiencies of Multicyclone-Variants ST16-L (top), ST16-S (center) and UC16 (bottom) with $\Delta p = 3000$ Pa.

Multiclones with 30 Cells

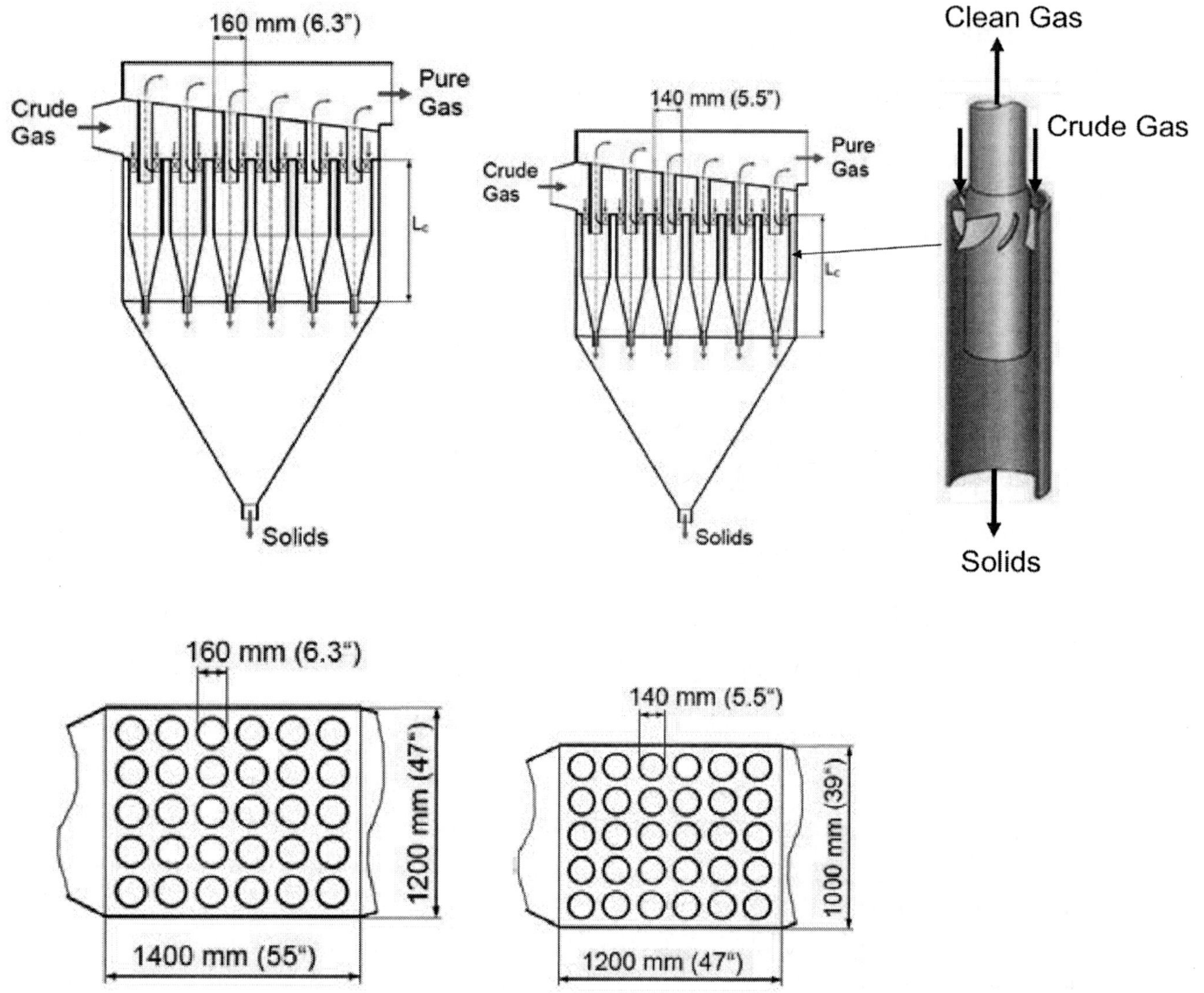

Fig. 7. Multicyclone-Variants ST30-L (left) and ST30-S (right) with 30 Swirl Tubes.

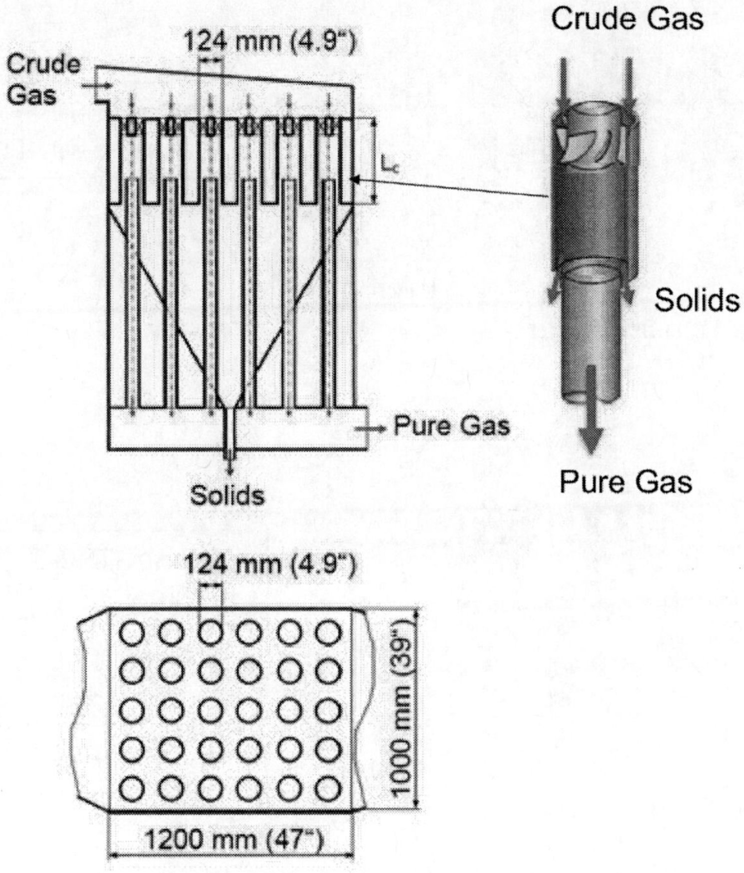

Fig. 8. Multicyclone-Variants ST30-L (left) and ST30-S (right) with 30 Uniflow Cyclones.

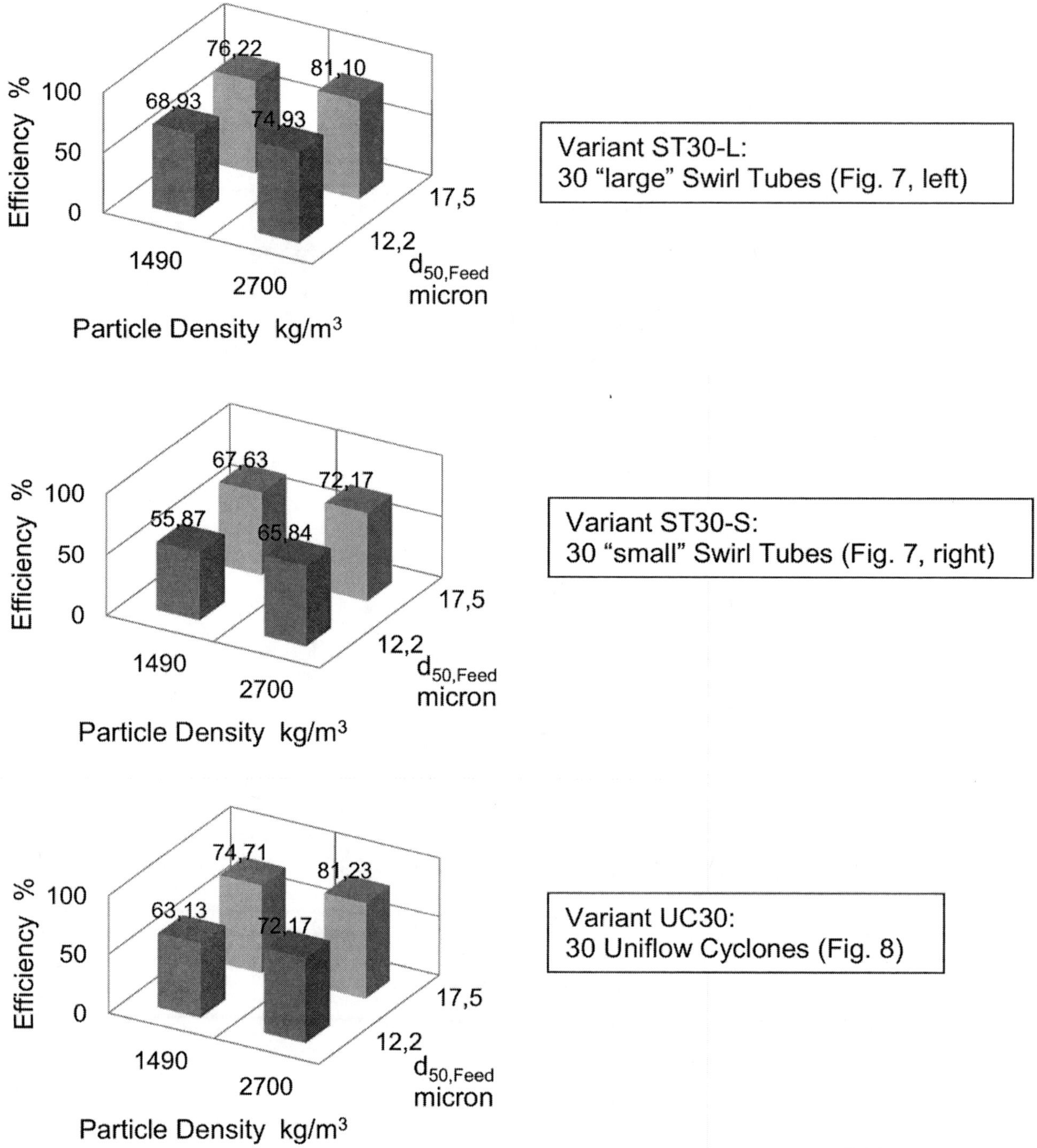

Fig. 9. Calculated Separation Efficiencies of Multicyclone-Variants ST30-L (top), ST30-S (center) and UC30 (bottom) with $\Delta p = 1000$ Pa.

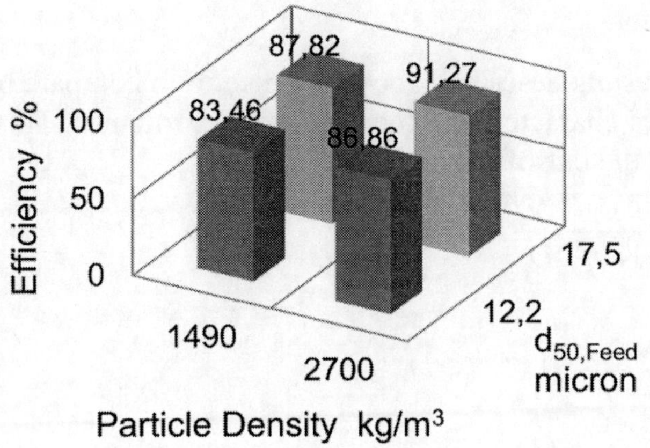

Variant ST30-L:
30 “large” Swirl Tubes (Fig. 7, left)

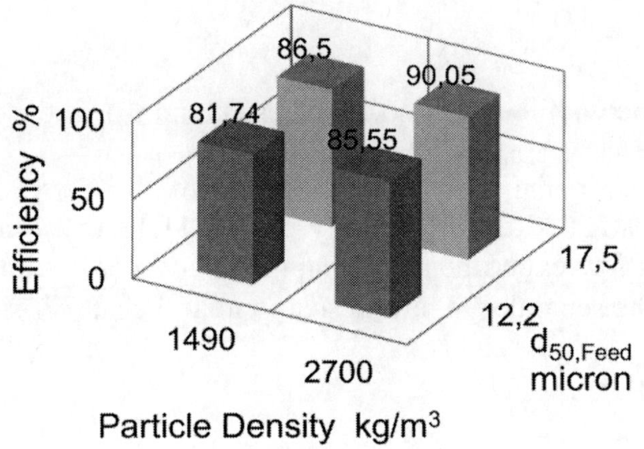

Variant ST30-S:
30 “small” Swirl Tubes (Fig. 7, right)

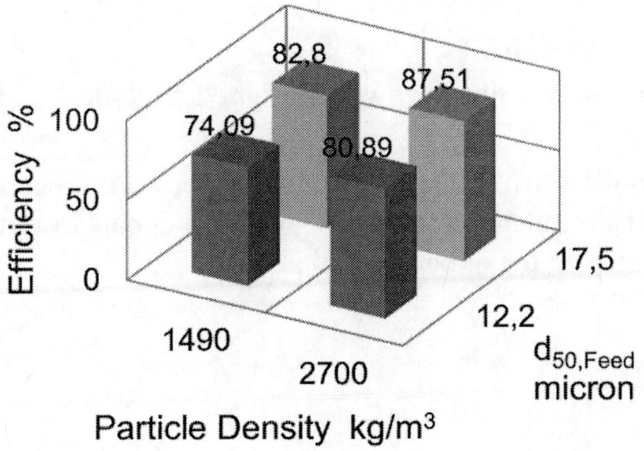

Variant UC30:
30 Uniflow Cyclones (Fig. 8)

Fig. 10. Calculated Separation Efficiencies of Multicyclone-Variants ST30-L (top), ST30-S (center) and UC30 (bottom) with $\Delta p = 3000$ Pa.

Comparison of the cut-off sizes

To obtain a better understanding of the results described above it is useful to compare both cyclone types based on dimensionless parameters characterizing the cyclone performance. This is done by applying the Stokes number of particles with the cut-off size

$$Stk(d_c) = \frac{\Delta\rho\, d_c^2\, v_z}{18\, \eta_g\, D_c} \qquad (2)$$

and the Euler number

$$Eu = \frac{\Delta p}{\frac{1}{2}\rho_g\, v_z^2} \qquad (3)$$

following a proposal described in (25). In eqs. (2) and (3)

$$v_z = (4/\pi)\, Q / D_c^{\,2} \qquad (4)$$

is the mean axial gas velocity through the cyclone with the cyclone diameter D_c and $\Delta\rho_g$ is the difference between the particle and the gas density.

Figs. 11 and 12 show the Euler number versus the Stokes number of the above described multiclones as well as of uniflow cyclone investigated experimentally in (6,9,11,13,15). Since grade-efficiency curves have been measured only in a few experiments, the cut-off sizes were estimated from the overall separation efficiencies η, assuming the separator to have a sharp cut at the cut-off size

$$Q_3(d_c) = 1 - \eta \qquad (5)$$

In addition the line resulting from the equation

$$Eu\,\sqrt{Stk(d_c)} = \sqrt{12} \qquad (6)$$

is plotted in Figs. 11 and 12, which provides a scaling rule for "well designed cyclones" at low solids concentrations according to (26).

As can be seen from Fig. 11 the data points of the calculated swirl tubes are close to the line resulting from eq. (6), whereas the data points of the calculated uniflow cyclones depart from that line at larger Stokes numbers beyond 0.01, in agreement with the experimental points.

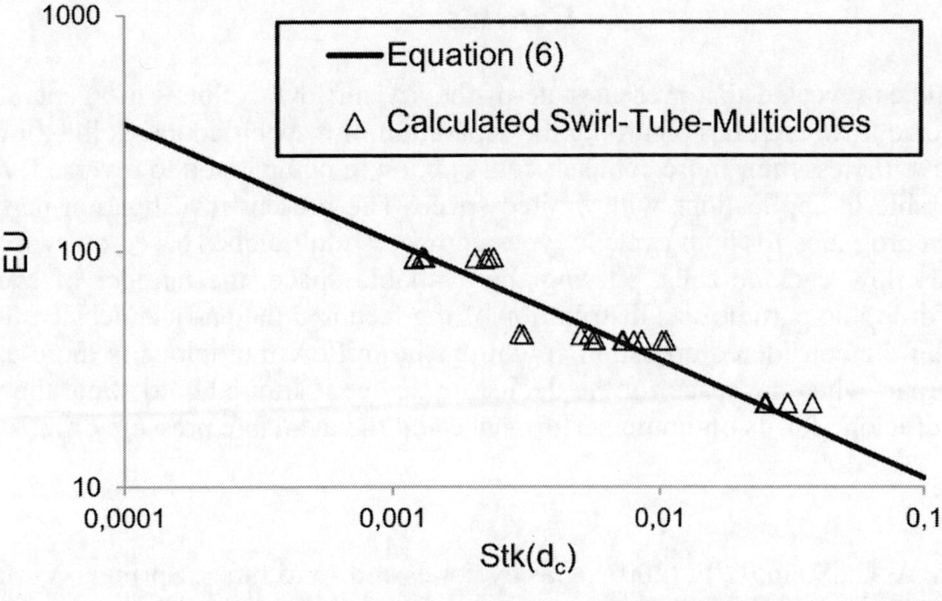

Fig. 11. Euler number *Eu* versus Stokes number *Stk*(d_c) showing calculated points for the investigated Swirl-Tube-Multiclones and the line for "well designed standard cyclones" according to the equation of Swarovski.

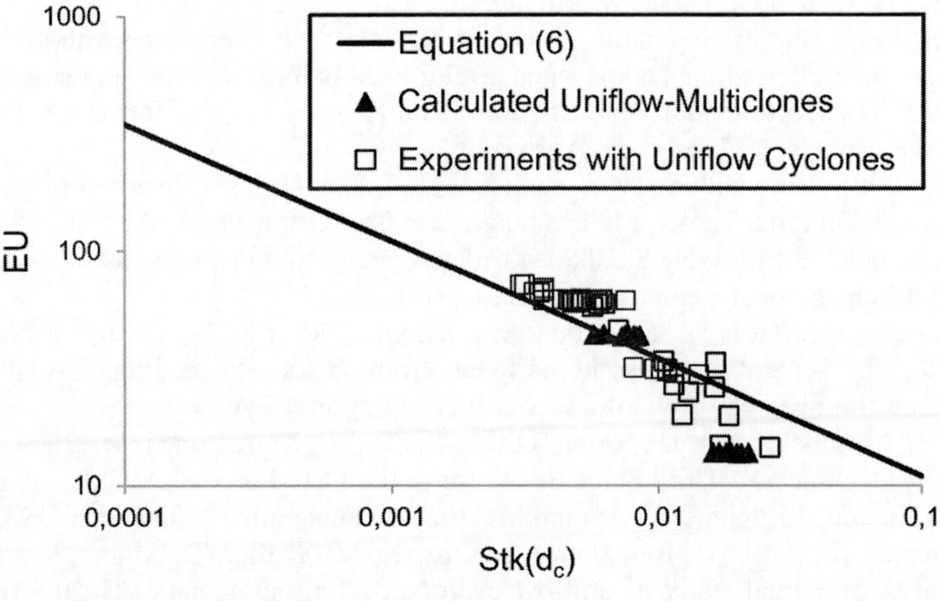

Fig. 12. Euler number *Eu* versus Stokes number *Stk*(d_c) showing calculated points for the investigated Uniflow-Multiclones and experimental points from uniflow cyclones together with the line for "well designed standard cyclones" according to the equation of Swarovski.

Conclusion

Recent studies revealed that present state-of-the-art uniflow cyclones are generally slightly less efficient than reverse flow cyclones regarding the separation of a specified gas-solids flow at a specified pressure drop. Nevertheless their more compact construction in comparison to reverse flow cyclones can make them preferable in applications with limited space. The present investigation performed by well proven calculation programs for both cyclone types compares multiclones based on swirl tube cells with those based on uniflow cyclone cells, varying the available space, the number of cyclone cells, the allowed pressure drop, the particle size distribution of the feed and the particle density. The calculations indicate that, under the considered operation conditions, a uniflow multiclone is more efficient than its swirl tube counterpart when the space for the dedusting device is limited onto about the volume needed by the uniflow multiclone for its optimum performance and the available pressure drop is low.

References

1. Hoffmann, A. C., Stein, L. E. (2007), Gas cyclones and swirl tubes. Springer-Verlag, 2nd Edition.
2. Krishnamurthy, S., Koves, W. (2013), Apparatuses and methods for gas-sold separations using cyclones. Uop Llc. Patent US-8419835 B2.
3. Gauthier, T. A., Briens, C. I., Bergougnou, M. A., Galtier, P. (1990),. Uniflow cyclone separation efficiency. Powder Technol. 62, 217-225.
4. Weng, M. (2002), Experimentelle und numerische Untersuchung von Gleichtromzyklonen, Ph.D. Thesis, RWTH Aachen University, Germany.
5. Muschelknautz, U (2002), Separation of coarse particles: New separator with low pressure drop. Proceedings of 7th International conference on fluidized bed technology, Niagara Falls, Canada.
6. Würtl, R. (2007), Experimental investigation of uniflow cyclone performance. Diploma Thesis, MCI Management Center Innsbruck, Austria.
7. Foidl, F. (2008), Entwicklung von Leitapparaten zur Druckverlustminimierung in Gleich-stromzyklonen. Diploma Thesis, MCI Management Center Innsbruck, Austria.
8. Leitner, A. (2009), Einfluß der Staubart auf den Trenngrad eines Gleichstromzyklons. Diploma Thesis, MCI Management Center Innsbruck, Austria.
9. Muschelknautz, U., Pattis, P. Reinalter, M., Kraxner, M. (2011), Design criteria of uniflow cyclones for the separation of solid particles from gases. Proceedings of 10th International conference on fluidized bed technology, Sunriver / Oregon, USA.
10. Kraxner, M., Muschelknautz, U., Karri, S.B.R., Cocco, R., Knowlton, T.M., 2011. Applicability of a Uniflow Cyclone as a Third Stage Separator in the FCC-Process. AIChE – American Institute of Chemical Engineers – Annual Meeting, Minneapolis / Minnesota, USA.
11. Muschelknautz, U., Pattis, P., Reinalter, M., Kraxner, M., Kofler, T., Aigner, M., Eichwalder, G. (2012), An experimental study of uniflow cyclone design parameters. AIChE-Annual Meeting, Pittburgh, USA.
12. Kraxner, M., Muschelknautz, U., Wechner, S., Ackermann, S., Greif, V., Bolda, J., Sommer, K. (2012), Influence of the inlet vane geomatrey on the uniflow cyclone performance. AIChE-Annual Meeting, Piitsburg, USA.
13. Muschelknautz, U., Kofler, T., Kraxner, M., Schretter, P., Pattis, P. (2012), An experimental investigation of uniflow cyclones with a spiral inlet. 62nd Canadian Chemical Engineering Conference, Vancouver, Canada.
14. Kraxner, M., Portenkirchner, M., Brugger, T., Muschelknautz, U. (2012) Effective flow rates for

uniflow cyclone performance. 62nd Canadian Chemical Engineering Conference, Vancouver, Canada.
15. Kofler, T., Kraxner, M., Pillei, M., Lanthaler, R., Ilchmann, C., Goller, R., Muschelknautz, U. (2013), Effect of solids loading on the uniflow cyclone performance at low solids concentrations. AIChE – Annual Meeting, San Francisco, USA.
16. Kraxner, M. (2013), Empirical evaluation of design criteria for uniflow cyclones in multicyclone boxes, PhD Thesis, TU Munich, Germany.
17. Muschelknautz, U. (2012), Analytical approach for calculating the separation efficiency of uniflow cyclones, Chem. Eng. Technol. 35(12), pp. 2099-2106.
18. Muschelknautz, U. (2015), Analytical approach for calculating the performance data of uniflow cycones. 22nd International Conference on Fluidized Bed Conversion, Turku, Finland.
19. VDI-Wärmeatlas Atlas (2018), Zyklone zum Abscheiden fester Partikel aus Gasen. 12th German Edition.
20. Muschelknautz, E., Greif, V. 1997. Cyclones and other gas-solids separators, in Circulating Fluidized Beds (Eds: J. R. Grace, A. A. Avidan, T. M. Knowlton), Blackie Acad. & Profess., London, 181-213.
21. Muschelknautz, U. (2016), Uniflow Cyclones compared with Standard Cyclones. Proceedings of FILTECH 2016.
22. Muschelknautz, U. (2017), Performance comparison of uniflow cyclones and standard cyclones. Proceedings of 12th International conference on fluidized bed technology, Krakow, pp. 421-428.
23. Greif, V. (1996), Reduzierung des Druckverlustes von Zyklonabscheidern durch Rückgewinnung der Drallenergie und Erweiterung der Grenzbeladungstheorie auf kleine und kleinste Staubbeladungen. VDI-Fortschr.-Ber., VDI-Verlag, Düsseldorf, Reihe 3, Nr. 470.
24. Roschek, D., Dereschewitz, M., Reckweg, F., Muschelknautz, U. (2015),. Design of Cyclone Improvements in two CFB Boilers and Operation Experience. 22nd International Conference on Fluidized Bed Conversion, Turku, Finland.
25. Peng, Hoffmann, Dries (2004), Separation Characteristics of Swirl-Tube Dust Separators, AIChE J. , 50, pp. 87-96.
26. Swarovski, L. (1986), "Solid-Gas-Separation" in *Gas Fluidization Technology*, D. Geldart, ed. New York, Wiley.

508021 Hydroclyclone Experiments and Design Equations for Produced Water Desanding Operation

Abstract:

Hydrocyclones are commonly used for liquid solid separations [1] involving high feed flow rates. The advantages of the use of hydrocyclones for the water treatment in oil extraction operations [2] has been proven, such as for example the attainment of produced water for water injection operation [3]. Nevertheless, little information are available in the literature for the design of hydrocyclone desanders related to water-solid system relevant to water treatment in produced water operations.

In this work, we present the experimental results on a series of tests carried out on a purposely-made pilot unit hosting a single hydro cyclone and able to work at water flow rates up to 2.5 L s^{-1}. The setup is able to set independently the feed flow rate and the underflow flow rate. Mixtures of water and Tri-hydrate Alumina particles (1 to 70 Î¼m size) up to 2000 ppm v/v of solid concentrations have been tested. Solid particle density, particle size distribution and shape makes the tested system very similar to real water sand mixtures in produced water desanding operation. These results have been used to find the proper hydrocyclone design equation such as those proposed by Coelho and Medronho [4]. In particular, we have found the proper correlations for the cyclone pressure drops, and the cyclone cut size as a function of the liquid flow rate and the water flow ratio between the underflow and the feed stream.

REFERENCES

[1] L. Svarovsky, Introduction to solid-liquid separation, in: Solid-Liquid Sep., Elsevier, 2001: pp. 1â€“29. doi:10.1016/B978-075064568-3/50025-0.

[2] J.C. Ditria, M.E. Hoyack, The Separation of Solids and Liquids With Hydrocyclone-Based Technology for Water Treatment and Crude Processing, in: SPE Asia Pacific Oil Gas Conf., Society of Petroleum Engineers, 1994. doi:10.2118/28815-MS.

[3] M.S. Choi, Hydrocyclone Produced Water Treatment for Offshore Developments, in: SPE Annu. Tech. Conf. Exhib., Society of Petroleum Engineers, 1990. doi:10.2118/20662-MS.

[4] M.A.. Coelho, R.. Medronho, A model for performance prediction of hydrocyclones, Chem. Eng. J. 84 (2001) 7â€“14. doi:10.1016/S1385-8947(00)00265-5.

Copyright © American Institute of Chemical Engineers. All rights reserved.

507304 A Hydrodynamic Study of Subway Grating and Disk and Donut Trays in a 0.6-m Diameter Fluidized Bed Stripper

Abstract:

Fluidized bed strippers are flowing fluidized beds where an up-flowing gas removes (strips out) a gaseous product entrained in a down-flowing stream of the emulsion phase. The stripper has to facilitate the transfer of the gas in the interstices of the emulsion phase and that adsorbed on the particles into the bubbles of the stripping gas. Major industrial applications of fluidized bed strippers are in fluid catalytic cracking (FCC) and fluid coking processes.

The counter-current contacting between the down-flowing emulsion phase stream and the up-flowing bubbles stream in fluidized bed strippers is enhanced by the use of an assembly of horizontal or inclined baffles. The baffles are important because in un-baffled fluidized beds, the movement of the bubbles is unrestricted. So, as they rise, interact, coalesce and grow in size they tend to move laterally toward the center or, in some instances, the bubbles can consolidate in a jet stream that bypasses significant portions of the bed in what has been termed gas bypassing. Therefore, open beds would give too low of a mass transfer to be effective strippers. Installing baffles enhances gas-solids interaction by controlling the bubble size and improving bubble distribution over the stripper cross section. As is typical for two-phase countercurrent flow unit operations, fluidized bed strippers can also suffer from flooding at relatively high solids mass fluxes.

Literature shows that disk and donut trays give excellent stripping efficiencies. They are among the most widely-used fluidized bed stripper internals in the fluid catalytic cracking (FCC) process. Disk and donut internals are, however, expensive and prone to flooding. This study compares the hydrodynamics of simple and cheaper subway grating trays with disk and donut trays in a 0.6-m-diameter fluidized-bed stripper. The stripper column was 7.6 m tall and the test solids were FCC catalyst particles containing 8% fines, less than 44 microns. Solids flowing out of the stripper were transferred by a 0.25-m-diameter, 3.4-m long vertical standpipe to a 15-m tall, 0.3-m-diameter riser that conveyed the solids back to the stripper via primary and secondary cyclones. The measurements performed in the study included differential pressure and local bubble properties.

Results showed that the subway grating stripper built a higher-pressure head, operated more smoothly without flooding over a wider range of operating gas velocities and solids fluxes compared to the disk and donut stripper. Literature studies showed, however, that the stripping efficiency of subway grating tray strippers is significantly lower than that of disk and donut strippers. Therefore, while hydrodynamics suggests a superior subway grating stripper, the future challenge will be to find ways of raising its stripping efficiency. Installing more subway grating trays in the same vertical height will most likely improve stripping.

Copyright © American Institute of Chemical Engineers. All rights reserved.

506869 Exceptional Intruder Sphere Sinking Due to Local Fluidization of Apparently Fixed Powder Bed

Abstract:

It is well known that a powder bed can be fluidized when the velocity of an airflow through the bottom of the bed is higher than the minimum fluidization velocity u_{mf}, and that the powder bed is a fixed bed when the air velocity is smaller than u_{mf}. The fluidized bed has liquid like properties such as density and viscosity. When an intruder sphere is put into the fluidized bed, the sphere floats or sinks based on the difference between the fluidized bed density and the sphere density. The float-sink can be utilized for density separation; the continuous separators are commercialized for waste treatments in Japan. Fundamentals of the fluidized bed, such as properties of air bubbles and convective flow of particles, have been widely investigated by many researchers. Generally, of course, the air velocity $u_0 > u_{mf}$ is employed for the fluidized bed research. We have been focusing on the spheres' float-sink in the fluidized bed. During the float-sink experiments, we accidentally found that spheres can sink even if u_0 is smaller than u_{mf}. Therefore, we investigated the sphere sinking in the powder bed for $u_0 < u_{mf}$. We varied the air velocity ratio u_0/u_{mf} and the density ration ρ_{sphere}/ρ_{bed} (sphere density/powder bed density). The spheres sinking was observed in the range $0.83 < u_0/u_{mf} < 1.0$, indicating that the powder bed behaves as a solid for $u_0/u_{mf} \leq 0.83$ and as an apparently fixed bed for $0.83 < u_0/u_{mf} < 1.0$. The final depth of sphere sinking increases with increasing the sphere density in the range $1.3 < \rho_{sphere}/\rho_{bed} < 2.6$. However, surprisingly, smaller density spheres sink "deeper" than larger density spheres in the range $1.0 < \rho_{sphere}/\rho_{bed} < 1.3$. We recorded the state of the top of the power bed by a video camera. When $\rho_{sphere}/\rho_{bed} < 1.3$, air bubbles are observed to percolate from the powder bed surface at the location of the sinking sphere, whereas for $\rho_{sphere}/\rho_{bed} \geq 1.3$, such percolation bubbles are not present. The rising air stream can cause *local fluidization* around the sphere when the sphere density is close to the powder bed density. As a result, air voids can then form underneath the sphere and migrate to the powder bed surface where they are observed as emerging percolation bubbles. As the sphere density ρ_{sphere}/ρ_{bed} decreases from 1.3 to 1.0, the decreasing weight of the sphere allows larger regions of local fluidization to develop. This creates larger voids that allow the sphere to sink deeper in spite of the decreasing sphere density.

Copyright © American Institute of Chemical Engineers. All rights reserved.

508575 Structured Flow of Bubbles in Pulsed Fluidized Beds: Pattern Stabilization and Propagation

Abstract:

Gas-solid fluidized bed reactors are widely applied in different industrial processes, such as chemical, environmental and pharmaceutical industries, where efficient contacts between solid and gas phases are critical. The solids mixing and transport process are highly dependent on the bubble dynamics. Nevertheless, the nonlinear collective behavior arising from the dissipative interparticle collisions, and the seemingly chaotic coalescence and the breakup of bubbles give rise to complex and unsteady bubbly flows which are hard to predict, challenging every aspect of the optimization, control, design, and scale-up of gas-solid fluidized bed reactors [1].

Over the past years, lots of efforts have been made to manipulate the bed hydrodynamics via introducing additional degrees of freedom. Some intrusive methods, such as fractal injectors and internal baffles, are imposed to suppress the chaos of bubble dynamics and channeling effects [2,3]. Among those, supplying a periodically pulsed gas flow has been demonstrated as an efficient non-intrusive method to structure gas-solid fluidized beds. When subjected to a periodic pulsed flow, gas-solid quasi-2D fluidized beds can emerge a structured flow of bubbles, where gas bubbles rise in a staggered fashion with a characteristic wavelength, forming regular hexagonal patterns. Under certain operating conditions, changing the inlet pulsed flow profile could simply customize the bubble size and its spatial distribution while maintaining the ordered structure [3,4].

Although pattern formation provides great potentials of engineering fluidized bed reactors, its fundamental aspects remain largely unknown. A few computational attempts have been conducted to model this phenomenon, but none of them have yet captured the regular pattern of bubbles convincingly [5,6]. Only recently, we reported the first successful simulation reproducing the dynamics of pattern formation with a CFD-DEM (computational fluid dynamics-discrete element method) modeling approach [7]. These numerical results provide valuable physical insights into the fundamentals of fluidized bed dynamics.

This contribution investigates the experimental observations on the formation and the stability of regular patterns in quasi-2D pulsed fluidized beds, subjected to various conditions. We discuss the response of pulsed bed dynamics to different pulsed flows, particles shape, static bed heights, and we also report a phase diagram demonstrating the favorable operating conditions for pattern formation. By comparing the experimental results with CFD-DEM computational modeling, we also study how the patterns propagate and being disturbed with bubbles rising towards the top surface. These results are valuable from a fundamental point and provide great insights into the scale-up applications as well.

[1] S. Pannala, M. Syamlal, T. J. O'Brien, Computational Gas-Solids Flows and Reacting Systems: Theory, Methods and Practice, IGI Global, Hershey, 2011.

[2] J.R. Van Ommen, J. Nijenhuis, M.-O. Coppens,Reshaping the structure of fluidized beds, Chem. Eng. Prog. (2009) 49-57.

[3] M.-O. Coppens, J.R. Van Ommen, Structuring chaotic fluidized beds, Chem. Eng. J. 96 (2003) 117-124.

[4] M.-O. Coppens, M.A. Regelink, C.M. van den Bleek, Pulsation induced transition from chaos to periodically ordered patterns in fluidised beds, Proceedings of the 4th World Congress on Particle Technology, Sydney, 2002.

[5] K. Wu, L. de Martín, L. Mazzei, M.-O. Coppens, Pattern formation in fluidized beds as a tool for model validation: A two-fluid model based study, Powder Technol. 295 (2016) 35-42.

[6] X. Wang, M. Rhodes, Pulsed fluidization—a DEM study of a fascinating phenomenon, Powder Technol. 159 (2005) 142-149.

[7] K. Wu, L. de Martín, M.-O. Coppens, Pattern formation in pulsed gas-solid fluidized beds – the role of granular solid mechanics, Chem. Eng. J. (2017) (in press).

Copyright © American Institute of Chemical Engineers. All rights reserved.

509583 Intermediate Regime in Particle-Fluid Flow

Abstract:

Simulations of the multi-phase flow in fluidised beds are presented. Some of the results rely on a model describing the solid stresses in particle laden flow. The model is shown to be valid for slow and intermediate flows and is developed following the theory developed by Tardos et al. (2003). The study extends previous work by Makkawi et al. (2006), where a steady one-dimensional particle flow model was used to demonstrate the smooth merge of quasi-static flow with â €œliquid-likeâ€ behaviour. The work aims at studying the solid stress in conjunction with a number of numerical codes. Validation is then undertaken using electrical capacitance tomography (ECT) measurements.

The time series data of solid fraction fluctuation, averaged within 14.5-18.5 cm above the distributor level, at 3 different fluidisation velocities are studied. In addition, comparisons are undertaken using different simulation software, namely OpenFOAM, Fluent and MFix.

The comparisons undertaken are used to detect which modelling framework and which software are more suitable to represent the behaviour of particle laden flow. The merge between flow regimes and the characteristics of the intermediate regime bring to the conclusion that a more complex constitutive law for the solid stresses is needed.

References

[1] G.I. Tardos, S. McNamara, I. Talu, Powder Technology. 2003;131(1):23-39.

[2] Y.T. Makkawi, P.C. Wright, R. Ocone, Powder Technology. 2006;163(1):69-79.

Copyright © American Institute of Chemical Engineers. All rights reserved.

508470 On the Use of Structured Adsorbents in Pressure and Temperature Swing Adsorption Processes

Abstract:

The use of strucutred adsorbents in PSA and TSA processes has been suggested for many decades. However, there has been limited commercialization. This is rather surprising since much has been patented and published on this topic. In this presentation an overview on the use of strucutred adsorbents in PSA and TSA processes will be given. Examples will be taken from commercial processes, the patent literature, journal articles and the authors' own experiences with strucutred adsorbents in their adsorption process laboratory.

Copyright © American Institute of Chemical Engineers. All rights reserved.

506468 A New Theoretical-Empirical Model for Cyclone Design

Abstract:

A New Theoretical-Empirical Method for Cyclone Design

One of the theoretical bases for many cyclone design methods is the work of E. Feifel. One of those who continued and improved upon Feifel's work was G. Miczek. Although some of Miczek's work was published (two in Czechoslovakia and one in Chem. Eng. in conjunction with C. Doerschlag), most was not, and has been held as proprietary by my current (and past) businesses for commercial benefit. After 40 years of work within this field, benefitting from Miczek's work, I wish to give credit to Miczek for his work by sharing it with the scientific and industrial communities.

The method is initially based upon the Vortex Theory described by Feifel and others. This describes a vortex of maximum tangential velocity at some radius that is a function of the over-all geometry of the cyclone. While the diameter of the cyclone itself is a variable that affects the location and magnitude of this "Core Vortex" velocity, it is, contrary to popular belief, by no means the most significant variable. The Core Vortex velocity magnitude and position is more strongly a function of the inlet size, arrangement and location, as well as the gas outlet diameter. The magnitude of the Core Vortex tangential velocity is not only the primary measure of the inertial forces available for particle separation but also the primary contributor to cyclone pressure drop.

At the highest level, there are two primary methods for increasing cyclone separation efficiency:

1. **Increase the centrifugal force** (defined as the opposite of centripetal force): i.e. increase the tangential velocity at a given radius or decrease the radius at a given tangential velocity of the flow stream that is transporting the particle.
2. **Provide for a greater amount of time** for the particle to be transported to the collection point before the flow stream exits the cyclone.

Miczek's contributions can be summarized as a method that determines what he calls a "Specific Flowrate" for any given cyclone. This Specific Flowrate should be more properly defined as Specific Velocity which is defined as the average one directional gas flow velocity of a cyclone at some defined standard conditions. These standard conditions include the pressure drop across the cyclone. This approach of comparing Specific Velocity of various cyclones at standardized conditions and pressure drop allows for a description of "cyclone quality" that considers the cyclone pressure drop, capacity, Core Vortex Velocity and radiuses well as a measure of the cyclone residence time.

While cyclones are relatively simple to construct, the accurate prediction of their performance remains quite challenging and difficult for most engineers. By combining Miczek's theoretical model with the empirical corrections he developed, an accurate predictive model for cyclone pressure drop and fraction efficiencies was achieved. We have been able to refine and improve the coefficients of Miczek's empirical corrections as we have successfully designed hundreds of cyclones used in commercial and research applications with minimal error in predicted pressure drop and fraction efficiencies. The intent of this paper is to describe in detail the theory, research it is based upon, and provide guidance for the continued refinement of the empirical coefficients by others who wish to utilize this method.

Copyright © American Institute of Chemical Engineers. All rights reserved.

508860 Particles in Contact: The Key Challenge in Solids Processing

Abstract:

One of the grand challenges in Particle Technology is how to characterize the contact mechanisms and contact forces between particles. Mechanical particle properties are the key parameters in discrete element models (DEM) which have been developed during the last decades and gradually reach "predictive power". Here and in many other aspects of Particle Technology, it is helpful to distinguish between process and materials functions. The process functions describe the effects of external parameters such as energy input, rotor revolution, machine geometry on the motion and stresses acting on particles in the process. As a result, the particles in the process or in the machine (such as mills, mixers or granulators) are stressed by these external forces and react accordingly depending on their material properties. In order to make DEM truly predictive, intrinsic material properties such as Young's modulus, hardness, plastic deformation or fracture toughness of the particles must be known quantitatively. For instance, we have shown how to model packing densities of glass beads as a function of particle size in dependency of intrinsic material properties. However, measured particle properties (and their distributions) are only known in a few rare cases.

The (nano-)mechanics of particles across all length scales down to the nanoscale are highly relevant for many applications including adhesion, friction, powder flow, comminution and tribology. Particular interesting new fields are additive manufacturing and 3D printing which "promise" a revolution in industrial manufacturing based on powder technologies. Nanomechanical elastic and inelastic particle properties have been measured systematically in a novel nanoindentation device installed within a scanning electron microscope to obtain simultaneously force-displacement information and optical images of deformed and broken particles. Quantitative data of intrinsic particle properties (and their respective distributions) are derived with high statistical relevance from force-displacement curves of single oxide, metal and polymer particles. The force-displacement of well-defined model particles can be used to obtain so far unknown distributions of stress energy and stress number distributions in mills and other solids processing devices. Finally, we show how novel 2D materials such as graphene, MoS_2 or BN can be produced by scalable mechanical delamination. Here, the challenge is also to characterize the delaminated nanosheets which is accomplished by advanced ultracentrifugation techniques. All these results are discussed in the context of unifying principles of product design and solids processing.

Copyright © American Institute of Chemical Engineers. All rights reserved.

508411 Impact Behavior of Microparticles on Component Surfaces Considering the Micro-Topography: Experiment and DEM Simulation

Abstract:

The Discrete Element Method (DEM) is widely used to simulate particulate processes. The walls of apparatuses and tolls in these simulations are often calculated as smooth surfaces, and the micro-topography, for example the surface roughness, is not considered. However, with decreasing particle size the influence of the surface asperities is becoming more and more important, because the deformation of single asperities and adhesion forces in the contact area can become dominant compared with the entire deformation and adhesion of the particle. On the micro scale the contacts with asperities take place often oblique whereas the impact on the macro scale is in normal direction to the wall surface. Furthermore there is no continuous contact area and the contact force during impact is distributed over a many single asperities. This leads to an additional energy dissipation and influences the rebound angle and rotation of the particle. Therefore it is necessary to describe the influence of the surface micro-topography on the particle impact behavior depending on the particle size.

In this work the impact behavior of microparticles with component surfaces was investigated by the simulation using DEM and experiments. The main focus was on the influence of the surface micro-topography on the rebound angle and the coefficient of restitution, which depends on the ratio between kinetic energies of the translational movement after and before particle impact.

The impact tests of single microparticles with different surfaces were performed. For these experiments a novel setup was developed. The collision of spherical particles with a diameter varied in the range of 100-700 μm was captured three-dimensionally with three high-speed cameras coupled with the microscope. The particle was dropped vertically to the surface to obtain a normal impact. The trajectory, angles and velocities of the particle were measured directly during the impact and rebound. The tests were performed with a polished and an unpolished titanium surface and smooth polystyrene microparticles. The experiments were repeated several times for each particle size and surface to measure the distribution of rebound angle and coefficient of restitution depending on the micro-topography.

To study the micromechanisms of the collision the DEM simulation was used. The titanium samples investigated by experiments were measured by Scanning Probe Microscopy (SPM) imaging with a nanoindenter (TI Premier, Hysitron) to obtain the micro-topography of surfaces. The SPM imaging was performed with a diamond Berkovich tip (with a tip radius of 100 nm) at a constant normal force of 2 μN. A total area of 2 mm x 2 mm was scanned by 40 x 40 consecutive areas with a size of 50 μm x 50 μm. The total area was reconstructed by merging the measured areas of the SPM imaging. A CAD model of this area with the resolved micro-topography was obtained by reverse engineering for the polished and unpolished surfaces. The point cloud of the surface obtained from the SPM imaging was meshed in an unstructured triangulated surface and implemented directly in the DEM simulation environment. The mesh resolution of the surfaces was varied to investigate the influence of the triangle size on the simulation results. With coarsening and refinement of the triangulated grid a mesh convergence could be obtained. Additionally an ideal smooth surface was simulated. The roughness of the polystyrene particles was neglected in the simulations, because it is by a factor of three to five smaller than the roughness of the investigated titanium surfaces. The required parameters for the contact model, like the coefficient of restitution and elastic modulus, were obtained by the impact tests and nanoindentation of the samples and particles. The impact point of the particles on the microstructured surface was randomly varied in the simulation to obtain the distribution of the parameters which describes the influence of the surface roughness. The simulation results for the different surfaces (unpolished, polished and ideal smooth) were compared with the experiments using the mean values and distributions of the coefficient of restitution and the rebound angle.

Copyright © American Institute of Chemical Engineers. All rights reserved.

508475 Toward Simulation-Based Design of Pharmaceutical Processes

Abstract:

Particle and fluid-particle simulations have been widely used to model a range of processes across many industries. This presentation will review application of simulations to common pharmaceutical processes including material transport and storage, blending, granulation, comminution, compaction, coating, and die filling. In many cases, simulations enable detailed analysis and flow visualization which provide insights into the effect of material properties and operating conditions promoting new designs and innovations and reducing inefficiencies and cutting costs associated with additional experimentation and scale-up studies.

Additionally, this presentation will review some computational models and techniques common to simulations involving particles. Recent advances in fluid-particle and particle models have significantly expanded the range of pharmaceutical systems and processes that can be described. Some of these advances include the modeling of particle mixes with wide variation in size, particle roughness, particle friction and particle cohesivity due to van der Waals or liquid bridging forces. Other model extensions include descriptions for particle aspericity that can also allow for particle breakage and flexibility. All of these complexities considerably affect the processing behavior of the particles. For example, surface moisture results in particle agglomeration, as is well known, but the particle aspect ratio determines the size and packing (solids volume fraction) of the agglomerates that are formed and the force required to convey or blend the particle mixture. And, when the solid volume fraction approaches the maximum packing density, the particle friction coefficient can be a more dominant factor than the liquid bridge force in determining the solid-phase shear stress.

Copyright © American Institute of Chemical Engineers. All rights reserved.

508976 Novel Technologies to Improve the Bioavailability, Content Uniformity and Manufacturing of Pharmaceuticals

Abstract:

Approximately 65% of all prescription drugs are manufactured as solid dosage forms, which includes tablets and capsules. For very potent drugs, the amount of Active Pharmaceutical Ingredient (API) in the solid dosage form can be as low as 0.1% by weight. This very low API loading poses one of the biggest challenges in pharmaceutical product development: the control of dose uniformity. Low API content variability in the blend are highly desired and strictly enforced by the U.S. Food and Drug Administration. Another challenge in product development is that about 40% of newly discovered drug compounds have poor solubility which subsequently affects their bioavailability. An important aspect of pharmaceutical process development is the final product cost. As pharmaceutical companies strive to develop more affordable drugs, any possible elimination of lengthy and expensive unit operations becomes commercially advantageous. One group of such unit operations is associated with the control of API attributes (size, size distribution, shape, bulk density, etc.). These unit operations can include crystallization and various milling steps. The need for control of API attributes is solely dictated by the drug product development and usually is associated with desired improvements in blend uniformity or release profile. Having a formulation process that can make these and other steps unnecessary can provide a large advantage to pharmaceutical companies. In this talk, we will discuss a manufacturing method for solid dosage forms using fluidized bed impregnation of APIs onto porous excipients (inactive filler powders) that can potentially address all of the above challenges in drug substance and product development.

Fluidized bed impregnation can be summarized as the combination of three distinct processes taking place simultaneously. These include fluidization of the porous carrier, spraying API solution within the bed, which penetrates the excipient due to capillary forces, and drying of the porous particles causing the API to be deposited within. In this talk, we will present results showing that this process can deliver powder with high blend uniformity and tablets/capsules with high content uniformity. In addition, we show that the final API loading is not limited by its solubility in the organic solvent. Our results also show that fluidized bed impregnation does not depend on the nature of the API but rather on the nature of the excipient used. The product is an excipient impregnated with API that has flow properties that depend on the excipient and not on the physical properties of the API. In addition, we show that the product that is produced can significantly increase the dissolution kinetics of poorly soluble APIs. Implementing fluidized bed impregnation in drug manufacturing could allow for significant cost savings due to elimination of several unit operations.

Copyright © American Institute of Chemical Engineers. All rights reserved.

513757 Particle Technology Impact on Pharmaceutical Product Performance and Process Design: Past, Present and Future

Abstract:

At the center of pharmaceutical product development lies the needs of patients. Each time a patient takes a drug, the patient expects consistent product quality and performance. These attributes are, in large part, determined at the product and process design stage. In recent years, particle technology has played a key role in advancing science-based approaches to product and process design shifting away from traditional empirical approaches. These science-based approaches drive efficiency and innovation in product development improving the pharmaceutical industryâ€™s ability to meet the patientâ €™s needs. Today particle technology applications span the drug development life cycle, from early stage materials characterization, to mid-stage product design and late stage process development and scale-up. This talk will present an industry perspective on how particle technology has transformed the product development paradigm. Real-world industrial examples will be presented. Current gaps and opportunities for future innovations will also be presented.

Copyright © American Institute of Chemical Engineers. All rights reserved.

507392 Particles Wall Coating Due to Electrostatic Charge Generation in Gas-Solid Fluidized Beds in Turbulent Versus Pre-Turbulent Flow Regimes

Abstract:

Continuous contacts amongst the particles and the particles and the fluidization vessel wall give rise to electrostatic charge generation in gas-solid fluidization processes. This might lead to significant operational challenges including fluidizing particles adhesion to reactor wall, particle agglomeration, and electrostatic discharges. One industry that has significantly suffered from negative effects of fluidized bed electrification is the petrochemical industry, where fluidized bed reactors are used for gas-phase ethylene polymerization to produce polyethylene. In such reactors, electrostatic charge generation results in catalyst and polyethylene particles to adhere to the reactor wall, causing a problem known as â€œsheetingâ€. The formation of particle sheets can require shutdown periods for clean-up which results in significant economic losses. To better understand the particles reactor wall coating due to electrostatics in such process, it is beneficial to conduct studies in relation to polyethylene reactor electrifications under industrially-relevant operating conditions of elevated pressures and turbulent flow regime. In this work, the effect of gas velocity on the degree of particles wall coating was studied with a specific focus on the transition to turbulent flow regime. Experiments were conducted in a pilot-scale fluidized bed operating under industrially-relevant conditions of a pressure of 2600 kPa (abs) and fluidizing gas velocities of 1.5, 3, and 5 times of U_{mf} (pre-turbulent regimes) and 7.5 times of U_{mf} (turbulent regime). Polyethylene particles directly received from a commercial reactor were fluidized in this work. Increasing the gas velocity and transitioning to the turbulent flow regime improved particle-wall contacts; and thus, augmented the extent of wall fouling, which indicates the increase in bed electrostatic charge generation. The amount of particles wall coating was approximately five times larger in turbulent flow regime (7.5 U_{mf}) in comparison to that for the lowest gas velocity examined in bubbling flow regime (1.5 U_{mf}). The particles coating on the column wall consisted of a thick bottom layer which extended to the expanded bed height, and a thin top layer which extended to the top of the column near the outlet. The particles net specific charge in the top and bottom layers did not vary with the increase in gas velocity. However, the net charge of these particles increased. The fine particles entrained from the bed had a net negative charge resulting in a net positive charge to be left behind in the bed contributing to the increase in the magnitude of wall fouling at higher gas velocities, especially in turbulent flow regime.

Copyright © American Institute of Chemical Engineers. All rights reserved.

507583 Process Intensification By Miniaturization of Fluidized Beds

Abstract:

Chemical micro-process technologies are relatively new concepts that offer much promise in terms of achieving process intensification by taking advantage of excellent mass and heat transfer characteristics of micro-structured devices. Recently there is a trend of miniaturization of fluidized bed systems to utilize further enhancement in mixing, heat and mass transfer provided by fluidized particles. However, there are some fundamental differences with classical macro-scale counterparts beds which need to be taken into account when designing these kinds of devices. The two major differences are the unavoidable wall effects due to small bed sizes and more importantly surface forces which start to be comparable or even dominate over gravitational force. These can be used to define the boundary between micro and macro-scale fluidization at 1 cm with a stricter limit at 1 mm in line with general microfluidics.
In terms of applications, the micro-fluidized bed holds promise in the already mentioned micro-process and micro-fluidic context. Besides fundamental studies on mixing and mass transfer enhancement, the current applications are mainly in high-throughput screening and testing of catalysts. The miniaturized fluidized bed also enables fast and reliable measurements of solid catalyzed reaction kinetics. Some more specific applications are in a low-cost device for detection of infectious bacteria, granulation screening of pharmaceutical tablet fillers, photocatalytic reactors and compact carbon dioxide HVAC units.

Acknowledgment

The authors acknowledge the support of the European Commission under the HORIZON 2020 Programme, for EU Project 680565 â€“ â€˜Intensified by Designâ€™ (IbD).

Copyright © American Institute of Chemical Engineers. All rights reserved.

507730 Scaling Analysis of Rectangular Spouted Bed Dynamics

Abstract:

2017 AIChE 8th World Congress on Particle Technology

Topic/Group: Fluidization & Multiphase Flow

Tentative Session: Fundamentals of Fluidization

Title: Scaling Analysis of Rectangular Spouted Bed Dynamics

Authors:

Steven L. Rowan: Steven.Rowan@netl.doe.gov

Jingsi Yang: Jingsi.Yang@netl.doe.gov

Ronald W. Breault: Ronald.Breault@netl.doe.gov

Key Words: Spouted Beds, Scaling, Cold Flow, Energy

Abstract:

Spouted bed reactors are used extensively in the chemical and food industries for the processing and drying of large, coarse particles that are not suitable for use in other multiphase flow systems such as fluidized beds. A thorough understanding of the dynamic performance of these units is critical when scaling up from small bench-scale units to larger production scale systems. An experimental study has been conducted to examine the hydrodynamic characteristics of rectangular spouted beds of geometric scales.

For this study, experimental data has been obtained using a small scale 4-inch x 1-inch rectangular spouted bed, as well as a larger 12-inch x 3-inch bed, using a variety of bed material, including 707 μm glass beads, 860 μm high density polyethylene, and 750 μm hollow alumina spheres. Hydrodynamic similarity between the two spouted bed reactors is explored via a statistical (skewness and kurtosis), frequency (power spectral density), and autocorrelation analysis of differential pressure data. In addition to analysis of differential pressure measurements, high speed video was recorded and used to determine the bed expansion ratio and the fountain height as a function of gas velocity. From the data obtained, correlations for various spouting regime transition velocities are proposed.

Copyright © American Institute of Chemical Engineers. All rights reserved.

509344 On the Mechanism of Anomalous Sphere Sinking in Apparently Fixed Particle Beds (Discrete Particle Simulation and Ultra-Fast MRI Measurement)

Abstract:

It is well known that a particle bed behaves as a fluid in the case it is aerated with gas velocity higher than the minimum fluidization velocity Umf. This is called as a fluidized bed and widely used in a large number of industrial processes. Under Umf, on the contrary, gravity force working on a particle overcomes fluid force and a particle bed behaves as a solid. Oshitani et al. (Phys. Rev. Lett., 116, 068001, 2016) reported anomalous sinking of intruder spheres in granular beds aerated under but close to Umf. They observed deep and slow sinking of spheres in the case density of sphere is close to the bulk density of the bed. They attributed this phenomena to local fluidization around the spheres while its detailed mechanism was not clear due to opaque nature of granular beds. In the present study, anomalous sinking and stopping mechanism of intruder spheres in an apparently fixed bed aerated just below Umf were explored and revealed in detail by means of discrete particle simulation and ultra-fast MRI measurement. Our numerical simulation and MRI measurement enable us to observe detailed flow dynamics and forces working on the spheres with sufficient spatial and temporal resolutions. Fluid force, contact force from surrounding particles and the gravity force working of the sphere are almost balanced in the case sphere density matches with the bulk density of a bed and small deviations from the equilibrium make slow sinking. The deviation is due to the detachment of small bubbles from the sphere surface. Bubble detachment from the sphere surface loosens local gas pressure gradient that supports the sphere weight and the sphere falls at the time.

Copyright © American Institute of Chemical Engineers. All rights reserved.

Flow Structure Study on Expanded Bed in Gas Fluidization of Fine Ellipsoids

*Jieqing Gan**, *Zongyan Zhou, Aibing Yu*

Laboratory for Simulation and Modeling of Particulate Systems, Department of Chemical Engineering, Monash University, VIC 3800, Australia

Abstract

Combined approach of computational fluid dynamics for gas phase and discrete element method for particles (CFD-DEM) is used to exhibit the formation of expanded bed in gas fluidisation of fine ellipsoids. The results illustrate that in the start-up stage, the bulk disconnects into two major bulks with a lowest contact force on the axial direction, especially for flat and elongated particles. The positions of the lowest point of contact force slightly change with time for oblate and prolate particles, but it moves smoothly from bottom to top for spheres. A slower gas introduction process has a mildest change in bed height (porosity). The fractures between major bulks gradually disappear when the cohesive forces become smaller. In the transition stage, when aspect ratio deviates from 1.0, the pressure drop fluctuation significantly increases. In the stable stage, when the aspect ratio decreases or increases from 1.0, the contact forces and van der Waals forces increases, and the particle-fluid forces become less uniformly distribution in the beds. Ellipsoidal particles undergo an orientation rearrangement during the expansion process.

Key words: CFD-DEM, expanded bed, fine ellipsoid, interparticle force.

Introduction

Geldart A (1973) particles, referred as 'aeratable', are characterized with an interval of gas flow beyond the point of minimum fluidization over which the bed expands in an apparently ideal manner, without bubbling. One focus on the flow regime is the stabilizing expanded bed, which has been a topic for many years. Criteria have been obtained for its stabilization in terms of interparticle force or particle-fluid forces [1, 2]. Another focus on expanded bed is the velocity range of the interval. Some correlations have been established to predict its range of the two demarcation point, that is the minimum fluidization velocity U_{mf} [3] and minimum bubbling velocity U_{mb} [4-10]. The presence of interparticle force was found to affect the bed bubbling [11, 12].

However, the formation process of the expanded bed is rarely reported. A few questions are still not clear. For example: what factors affect the formation process? How are the expanded bed formed? What are the microdynamics (e.g. particle velocity, orientation and interparticle forces) inside the bed at different stages of expanded bed formation? An insight into these questions could be useful for better understanding of the macroscopic formation of expanded beds. However, experiments experience difficulties in capture the instant information at particle scale, such as, particle contacts, orientation and forces. CFD-DEM becomes an alternative in the study of particle-fluid flow fundamentals. In this approach, the solid phase is based on the DEM, and gas phase is treated as a continuum phase in a similar way to the two fluid model. It has been used in the study of gas-fluidized beds of fine spheres [1, 13] and ellipsoids [13-15].

Although most particles in reality are non-spherical, only a few efforts had been made to investigate the effect of particle shape on the flow behaviour, and the role of particle forces on the formation of expanded of fine powders under the effect of particle shape is also not fully understand. This paper aims at investigating the formation process of expanded bed of fine ellipsoids using CFD-DEM method. The bed structure changes in three typical formation stages are discussed in terms of flow pattern, pressure drop, porosity, particle velocity, particle orientation and force ratio.

* Corresponding author. Email: jieqing.gan@monash.edu; Tel: 61 3 9905 0850.

Model description and simulation conditions

The governing equations for the translational and rotational motion of particle i with radius R_i, mass m_i, and moment of inertia I_i can be written as

$$m_i \frac{d\mathbf{V}_i}{dt} = \mathbf{f}_{f,i} + \sum_{j=1}^{k_c} (\mathbf{f}_{c,ij} + \mathbf{f}_{d,ij} + \mathbf{f}_{v,ij}) + m_i \mathbf{g} \quad (1)$$

$$I_i \frac{d\boldsymbol{\omega}_i}{dt} = \sum_{j=1}^{k_c} (\mathbf{M}_{t,ij} + \mathbf{M}_{r,ij} + \mathbf{M}_{n,ij}) \quad (2)$$

where $\mathbf{v}_i$ and $\boldsymbol{\omega}_i$ are the translational and angular velocities of the particle, respectively, and k_c is the number of particles in interaction with the particle. The forces involved are: particle-fluid interaction force $\mathbf{f}_{f,i}$, the gravitational force $m_i\mathbf{g}$, and interparticle forces between particles, which include elastic force $\mathbf{f}_{c,ij}$, viscous damping force $\mathbf{f}_{d,ij}$ and van der waals force $\mathbf{f}_{v,ij}$. These interparticle forces can be resolved into the normal and tangential components at a contact point. The torque acting on particle i by particle j includes: tangential torque $\mathbf{M}_{t,ij}$, and rolling friction torque $\mathbf{M}_{r,ij}$ and an addition torque $\mathbf{M}_{n,ij}$ because the normal contact force and van der waals force don't necessarily pass through the particle centre.

The governing equations for continuum fluid phase is given as

$$\frac{\partial \varepsilon_f}{\partial t} + \nabla \cdot (\varepsilon_f \mathbf{u}) = 0 \quad (3)$$

$$\frac{\partial (\rho_f \varepsilon_f \mathbf{u})}{\partial t} + \nabla \cdot (\rho_f \varepsilon_f \mathbf{u}\mathbf{u}) = -\nabla p - \mathbf{F}_{fp} + \nabla \cdot \varepsilon_f \boldsymbol{\tau} + \rho_f \varepsilon_f \mathbf{g} \quad (4)$$

where $\mathbf{u}$, ρ_f, p and $\mathbf{F}_{fp} = (\sum_{i=1}^{k_V} \mathbf{f}_{f,i}) / \Delta V$ are the fluid velocity, density, pressure and volumetric fluid-particle interaction force, respectively. $\boldsymbol{\tau}$ and ε_f are the fluid viscous stress tensor and porosity, respectively. The equations used to calculate the particle-particle interaction forces and torques and particle-fluid force can be found elsewhere [13-15].

The simulation begins with the poured packing of ellipsoidal particles in a rectangular box. Then gas is introduced from the bed bottom. The fluidization process is considered to have reached the pseudo-stable state when the bed pressure drop fluctuates around fixed values. Spheroids with aspect ratios varying from 0.25 to 3.5 are used in this work to produce shapes from oblate to prolate spheroids. Spheroids have three principal diameters. Let a be the principal radius in the polar direction, and b and c the principal radii in the equatorial plane. Here, the particle shape factor, namely aspect ratio α is defined as, $\alpha = a/c$. The particle properties used in the simulation are listed in Table 1.

Table 1. Simulation settings for particles and bed.

Parameters	Value
Particle size d_v, μm	50
Particle density ρ_p, kg/m^3	1349.5
Particle shape, [-]	Oblate(α<1), sphere(α=1), prolate (α>1)
Particle aspect ratio, [-]	0.25, 0.5, 0.75, 1.0, 1.5, 2.5, 3.0, 3.5
Number of particles N, [-]	18000
Bed size, [-]	$60d_p \times 480d_p$
CFD Mesh, [-]	20×160
Young's modulus E, N·m^2	1.0×10^7
Poisson's ratio ν, [-]	0.30
Friction coefficient μ, [-]	0.3
Rolling friction coefficient μ_r, d_v	0.002
Normal damping coefficient γ_n, [-]	2.0×10^{-5}
Hamaker constant H_a, J	0, 1.05×10^{-21}, 2.1×10^{-21}, 4.2×10^{-21}
Cut off distance h_{min}, nm	1, 5, 10

Results and discussion

When the gas velocity surpasses the U_{mf}, the bed is in expanded bed regime. The expanding process can divided into three stages, namely, start-up (corresponding to the increase in porosity at the beginning), transition (defined as the range of sharp descending in porosity), and stable (where all forces and porosity keep stable with time) [16].

Strat-up stage

From previous studies on spheres [1, 17] and ellipsoids [14], in the ascending process of the start-up stage, particles in the bed will disconnect into two bulks. To find out the reason, the effects of gas introduction method and particle cohesion are studied.

In our previous study [14], the gas is directly introduced with a constant gas velocity of U_g, but in most of the fluidization experiment, gas is gradually introduced to a specified gas velocity U_g. In present work, three gas introduction methods are studied: constant gas velocity introduction (case (a)), fast increasing gas velocity from U_{mf} to U_g within 0.25s (case (b)), and slowly introduce gas from velocity of U_{mf} to U_g within 0.75s (case (c)). The flow pattern, gas velocity and porosity (calculated from bed height) changes are plotted in Fig. 1. From Figs. 1(a), (b) and (c), it can be seen that for the three cases, there are fractures in the particle bulk. As shown in Fig. 1(d), case (a) the firstly to exhibit a highest porosity peak (corresponding to highest bed height), the gas does not have time to permeate the upper bed before they are lifted up by the bottom gas. The case (c) is the latest to show the lowest porosity peak as it reaches U_g in the latest, and the gas has enough time the permeate the upper bed and go through the whole bed. However, the finial stable porosity is quite close for the all three cases. Therefore, the gas introduction mode cannot eliminate the break-up of particle bulk, but a slower gas introduction process has a mildest change in bed height (porosity). In the following discussion, we still use the constant gas velocity introduction mode.

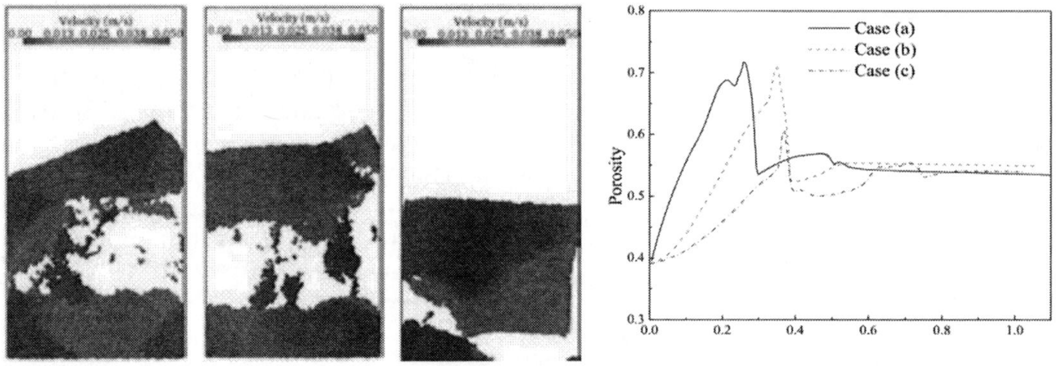

(a) (b) (c) (d)

Fig. 1 Flow patterns in the start-up stage of bed expansion for different gas introduction methods (0.2-0.4s). (a) Instant U_g, (b) Increase to U_g within 0.25s, (c) Increase to U_g within 0.75s, and (d) the variation of porosity with time. (Coloured by particle velocity)

The flow pattern of different Hamaker constant H_a are also compared and shown in Fig. 2. It can be seen that, with the decrease of H_a, the fracture between two bulks becomes obscure and even disappear when there is no van der Waals force. This indicates that the root reason for the formation of the fracture in the expansion of ellipsoids that the cohesive force on ellipsoids are much higher than the spheres, as was shown in Fig. 2.

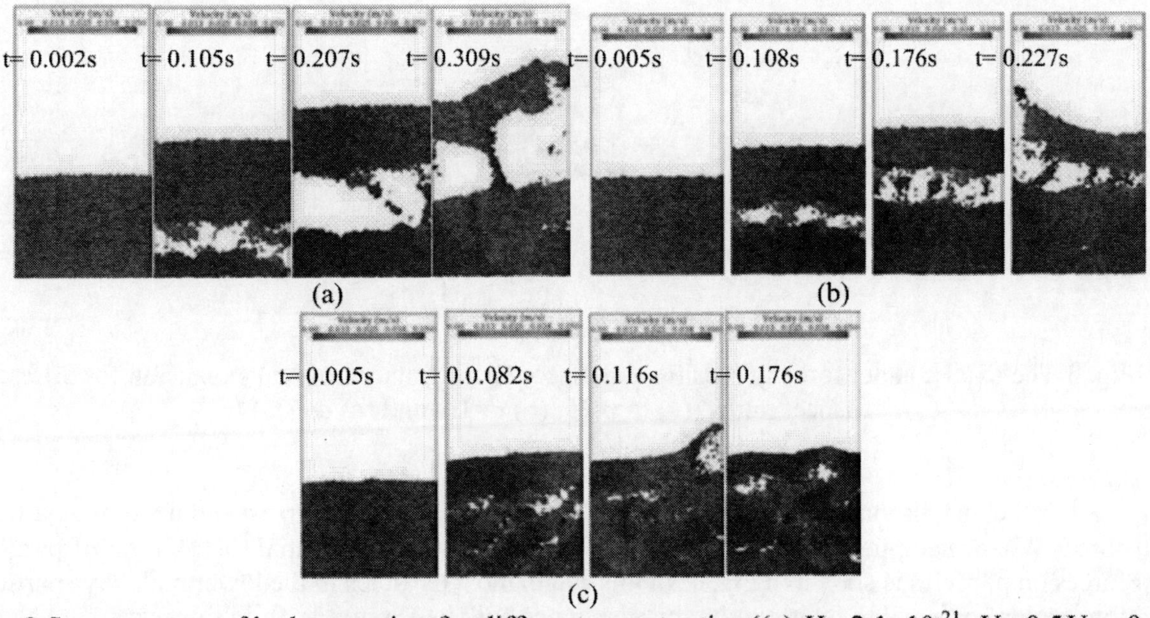

Fig. 2 Start-up stage of bed expansion for different aspect ratios ((a) H_a=2.1×10^{-21}, U_g=0.5U_{mb}=0.004 m/s, (b) H_a=1.05×10^{-21}, U_g=0.5U_{mb}=0.0035 m/s, and (c) H_a=0.0, U_g=0.0035 m/s, α=0.25, U_g=0.004 m/s, coloured by particle velocity)

The variation axial layer-averaged contact force with bed depth at different time in the start-up stage for different aspect ratios is shown in Fig. 3. At the beginning of gas introduction into the beds, the axial contact force increases to the maximum value and then decrease with the bed depth. This is because on one hand, the initial close packed bed will add weight to the underneath particles, and on the other hand, the blowing gas leads to looser bed bottom. When the beds are loose enough, the bulk disconnects, so there shows up a lowest point on the curves. However, there are obvious differences in the variation of lowest contact force with time for different aspect ratios. More specifically, for ellipsoids, with the time increases, the contact forces at the lowest point decreases which implying more weaker contacts, but the position of the lowest point slowly change with time. From Figs. 3(a) and (c), the upper bulk has much higher coordination number (CN) than the bottom bulk. For spheres (Fig. 3(b)), however, with the time increases, the lowest contact force position move from bottom to top, which means that the disconnect position move upwards as the particles in the upper bulk continuously fall to the bottom bulk, giving rise to a smooth expansion. Moreover, the contact forces in the bottom also increases due to the increasing height in the bottom bulk. From the inset of Fig. 3(b), the coordination number in the upper and bottom are quite close.

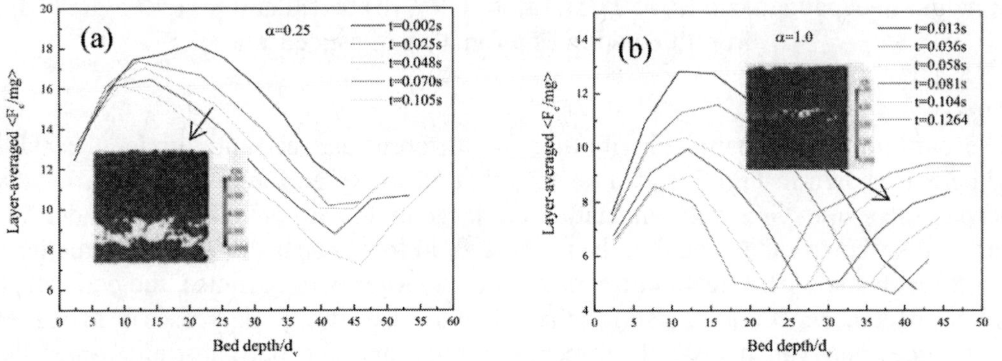

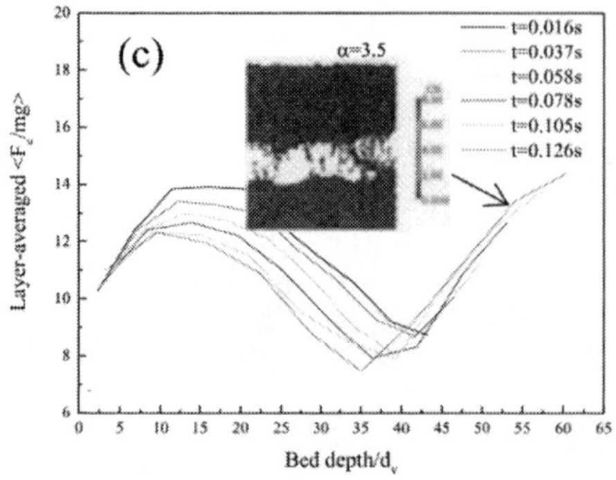

Fig. 3 The axial contact force in loosing and disconnecting stages of bed expansion for different aspect ratios ((a) α=0.25, (b) α=1.0 and (c) α=3.5)

Transition stage

After the bed disconnect into two bulks, the upper bulk continues to ascend under the particle-fluid force. When the upper bulk reaches the highest bed height, the spatial distribution of particle-fluid forces on particles is shown in Fig. 4. In this stage, most particles in the lower bulk have particle-fluid forces close or equal to their gravity, but the upper bulk have particle-fluid forces cannot balance their gravity which makes them falling to the bottom, so most particles have velocities in the downward direction. As shown in Fig. 4(d), when particles in the upper bulk fall to the bottom, there is a pressure drop fluctuation in the bed. For spheres, due to the relative continuous expanding process, the pressure drop fluctuation is quite small, but when aspect ratio deviates from 1.0, the pressure drop fluctuation significantly increases.

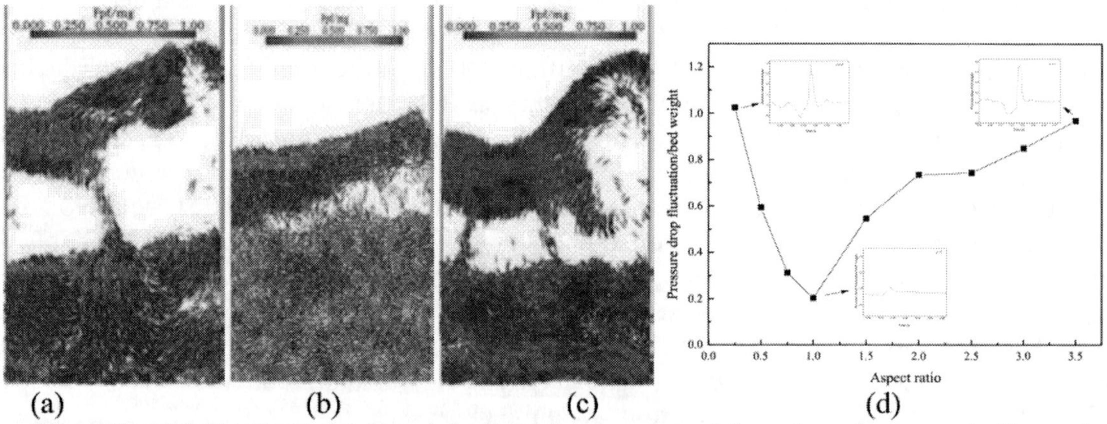

Fig. 4 Spatial distribution of particle-fluid forces on particles when the upper bulk reaches the highest height in expanded bed (d_p=50 μm, (a) α= 0.25, (b) α=1.0 and (c) α=3.5) and (d) pressure drop fluctuation as a function of aspect ratio

Stable stage

The particle forces spatial distribution for different aspect ratios under the stable state in expanded beds is shown in Fig. 5. It can be seen that all the three forces vary significantly with bed space and particle shape. Generally, in spatial, the large and small contact and van der Waals forces are uniformly distributed in the bed, but the particle-fluid forces near the voids are smaller than inside the bulk. Although some particle-fluid forces are much larger than particles, the bed can still be kept in stable. For different aspect ratio, it shows that when the aspect ratio decreases or increases from 1.0, the contact forces and van der Waals forces increases, and the particle-fluid forces become less uniformly distribution in the beds.

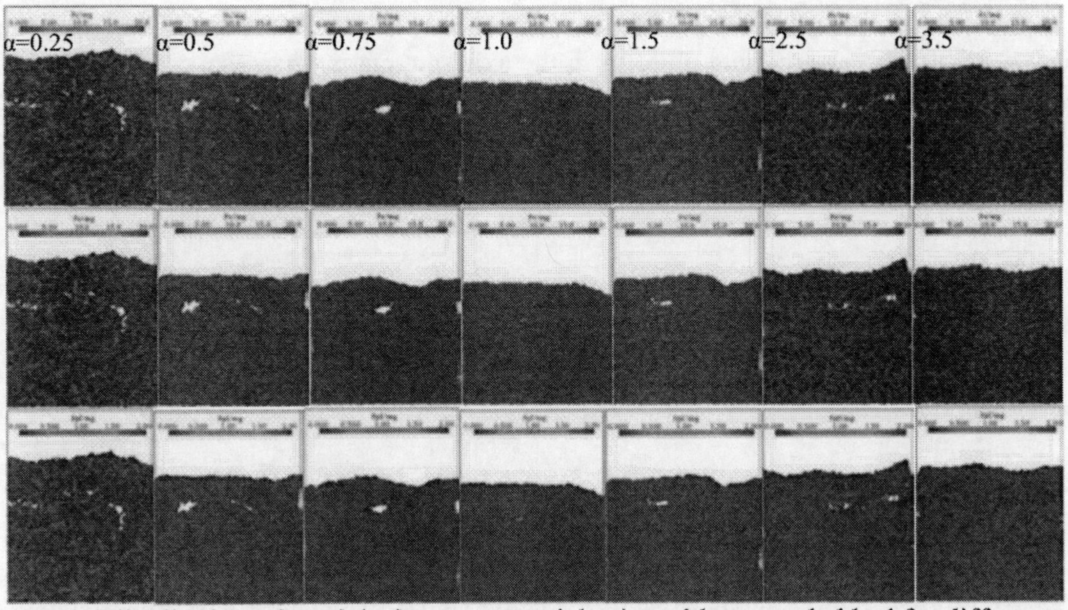

Fig. 5 Spatial distribution of particle forces on particles in stable expanded bed for different aspect ratios (top: $\mathbf{f}_c$/mg, middle: $\mathbf{f}_{vdw}$/mg, and bottom: $\mathbf{f}_{pf}$/mg).

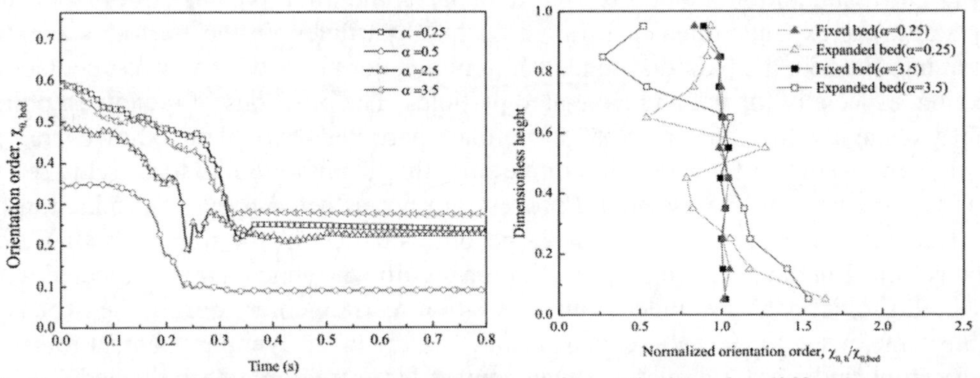

Fig. 6 (left) Variation of bed-averaged orientation order with time for different aspect ratios, and (right) axial distribution of the average orientation order in fixed and expanded beds (d_p*=50 μm*, fixed bed: U_g*=0.5* U_{mf} and expanded bed: U_g*=0.5* U_{mb})

From the flow patterns mention above, ellipsoidal particles undergo a position rearrangement during the expanding process, and meanwhile there are also changes in particle orientation. Fig. 6(a) shows that variation of orientation order with time for different aspect ratios. Here orientation order χ_θ is defined according to ref. [18, 19], and it becomes zero when the particles are randomly oriented and 1 when all particles lie in the horizontal plane. High orientation order is observed in the packed bed for different aspect ratios. With the introduction of gas, the particles orientation order gradually decreases and finally levels off at the stable stages. For different aspect ratios, the more non-spherical, the higher the orientation order is observed. The axial distribution of the average orientation order in fixed and expanded beds is compared and shown in Fig. 6(b). It can be seen that particles are axially uniform in the fixed beds. However, large orientation order fluctuations show up in the axial direction in the expanded beds, which proves that particles undergo orientation rearrangement during this process.

From the Fig. 5, although particle forces vary spatially, the beds are in force balance to keep a stable expanded bed. In our previous study [15], we have confirmed that the force-balance criterion is valid for the ellipsoidal particles. That is at stable stage, the van der Waals forces balance the particle contact forces, and the particle-fluid forces balance the weight of the bed. In this work, the axial contact forces in the stable expanded beds are studied, and shown in Fig. 7. It can be seen that in stable expanded beds, the forces are quite uniform in the axial direction when aspect ratio is close to 1.0. However, when aspect ratio deviates from 1.0, the layer-averaged contact forces fluctuate in the axial direction, and there is a relative lower contact forces region within the bed. This indicates that for flat or elongated particles, although the two bulks join together at the stable stage, there are still in loose contact near the joint.

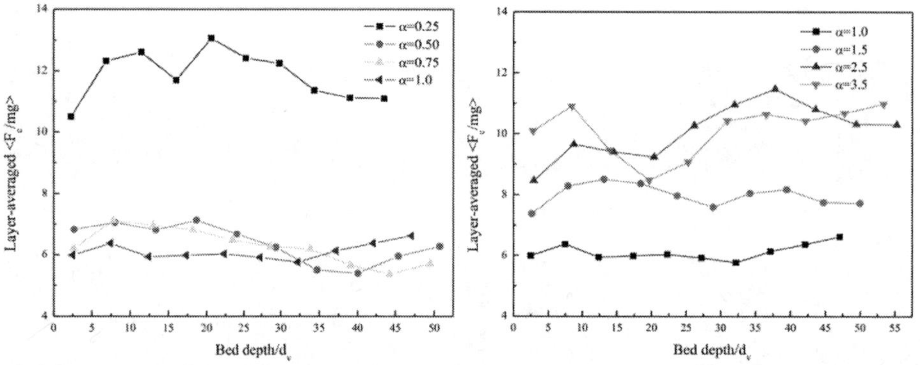

Fig. 7 Axial force ratio in stable expanded beds for (left) oblate and (right) prolate spheroids.

Conclusion

CFD-DEM simulation method is used to analysis the three typical stages (start-up, transition and stable stages) of expanded bed with fine ellipsoidal particles. In the start-up stage, with the gas introduction into the bed, the bulk disconnects into two major bulks with a lowest contact force on the axial direction, especially for flat and elongated particles. The positions of the lowest point of contact force slightly change with time for oblate and prolate particles. For spheres, however, with the time increases, the lowest contact force position move smoothly from bottom to top, giving rise to a smooth expansion. The fractures in the ascending process are not affected by gas introduction method, but gradually disappear when the cohesive forces become smaller. In the transition stage, when aspect ratio deviates from 1.0, the pressure drop fluctuation significantly increases. Compared with the initial packed bed, ellipsoidal particles undergo an orientation rearrangement during the expansion process. In the stable stage, when aspect ratio deviates from 1.0, the layer-averaged contact forces fluctuate in the axial direction, and there is a relative lower contact forces region within the bed.

Acknowledgment

The authors are grateful to the ARC Hub for Computational Particle Technology for the financial support of this work, and the NCI National Facility for the support in computation.

References

[1] Q.F. Hou, Z.Y. Zhou, A.B. Yu. Micromechanical modeling and analysis of different flow regimes in gas fluidization. Chemical Engineering Science. 2012;84(0):449-68.

[2] B.H. Xu, A.B. Yu. Numerical simulation of the gas-solid flow in a fluidized bed by combining discrete particle method with computational fluid dynamics. Chemical Engineering Science. 1997;52(16):2785-809.

[3] K. Suksankraisorn, S.Patumsawad, B. Fungtammasan. Prediction of the minimum fluidization velocity from correlations: An observation. Asian J Energy Environ. 2001;2(2):145-54.

[4] M. ŚciĄżko, J. Bandrowski. Effect of pressure on the minimum bubbling velocity of polydisperse materials. Chemical Engineering Science. 1985;40(10):1861-9.

[5] K.V. Jacob, A.W. Weimer. High-pressure particulate expansion and minimum bubbling of fine carbon powders. AIChE Journal. 1987;33(10):1698-706.

[6] J.M. Valverde, M.A.S. Quintanilla, A. Castellanos, P. Mills. Experimental study on the dynamics of gas-fluidized beds. Physical Review E. 2003;67(1):016303.

[7] M. Ye, M.A. van der Hoef, J.A.M. Kuipers. The effects of particle and gas properties on the fluidization of Geldart A particles. Chemical Engineering Science. 2005;60(16):4567-80.

[8] R. Girimonte, B. Formisani. The minimum bubbling velocity of fluidized beds operating at high temperature. Powder Technology. 2009;189(1):74-81.

[9] J. Wang, M.A. van der Hoef, J.A.M. Kuipers. CFD study of the minimum bubbling velocity of Geldart A particles in gas-fluidized beds. Chemical Engineering Science. 2010;65(12):3772-85.

[10] S. Rapagna, P.U. Foscolo, L.G. Gibilaro. The influence of temperature on the quality of gas fluidization. International Journal of Multiphase Flow. 1994;20(2):305-13.

[11] J.A. Agbim, A.W. Nienow, P.N. Rowe. Inter-particle forces that suppress bubbling in gas fluidised beds. Chemical Engineering Science. 1971 8//;26(8):1293-4.

[12] J.K. Pandit, X.S. Wang, M. Rhodes. Influence of interparticle cohesive force on fluidized bed Behaviour by DEM simulation. Third International Conference on CFD in the Minerals and Process Industries; 2003 10-12 December 2003; CSIRO,Melbourne, Australia; 2003.

[13] Z.Y. Zhou, D. Pinson, R.P. Zou, A.B. Yu. Discrete particle simulation of gas fluidization of ellipsoidal particles. Chemical Engineering Science. 2011;66(23):6128-45.

[14] J.Q. Gan, Z.Y. Zhou, A.B. Yu. CFD–DEM modeling of gas fluidization of fine ellipsoidal particles. AIChE Journal. 2016;62(1):62-77.

[15] J.Q. Gan, Z.Y. Zhou, A.B. Yu. Micromechanical analysis of flow behaviour of fine ellipsoids in gas fluidization. Chemical Engineering Science. 2017 5/18/;163:11-26.

[16] B.H. Xu, A.B. Yu. A new stability criterion for bed expansion of gas fluidisation. *4th world congress on particle technology*. Sydney, Australia. 2002:189.

[17] M.J. Rhodes, X.S. Wang, M. Nguyen, P. Stewart, K. Liffman. Onset of cohesive behaviour in gas fluidized beds: a numerical study using DEM simulation. Chem Eng Sci. 2001 7//;56(14):4433-8.

[18] G.W. Delaney, J.E. Hilton, P.W. Cleary. Defining random loose packing for nonspherical grains. Physical Review E. 2011;83(5):051305.

[19] B.J. Buchalter, R.M. Bradley. Orientational Order in Amorphous Packings of Ellipsoids. EPL (Europhysics Letters). 1994;26(3):159.

512547 Keynote Presentation: The Advancement of Fluid/Particle Separation for Environmental Protection

Abstract:

Environmental and health protection efforts continue to be key challenges and opportunities for the scientific community. Whether the challenge is air pollution caused by the particulate matter (e.g. PM2.5), fine particle removal for wastewater discharge/recycling, or sufficient dewatering of a solid for disposal, fluid/particle separation technologies have been playing key roles and evolving over the years. Recent technology developments and the future outlook are reviewed in this presentation. For clean air, filter media development is the major focus. Advanced filtration media allow for less fuel consumption in energy generation and water purification process and, as a result, reduces emissions.

For effluent water, improvements to traditional solid/liquid separation technologies, such as clarifier design and the use of membrane filters, have allowed industries to meet more stringent regulatory requirements. In the area of solid waste reduction, innovative technologies are also being developed. A new chemical amendment for difficult to dewater tailings is highlighted as an example.

Copyright © American Institute of Chemical Engineers. All rights reserved.

506343 Atomic Scale Engineering of Polymer Fibers for Enhanced Filtration Performance

Abstract:

Manipulation of atomic scale structures of molecules in fibers provides a method to control and modify material properties, including short range forces near fiber surfaces, for practical applications. In this work mats of small PVDF electrospun fibers are polarized. The fiber structure is observed on a stabilized high resolution SEM with indications of the atomic structure of the fluorine atoms as bright spots on the SEM image. The polarized mats are evaluated for surface charge and tested for performance in the capture of nanoscale salt particles from flowing air stream.

Copyright © American Institute of Chemical Engineers. All rights reserved.

507262 Wet Classification of Fine Particles Using Crossflow Filtration

Abstract:

1.Introduction

Due to the increasing use of particulate products with a fine and narrow particle size distribution, processes are needed to generate such particle systems. Particles below 10 μm are used for pigments, polishing compounds or pharmaceutics. Especially for these products, it is necessary that not a single coarse particle occurs. Conventional production processes like precipitation, crystallization or comminution of coarse particles cannot guarantee that all particles are smaller than a few μm or have a narrow particle size distribution at a small size. Additionally, conventional wet classification processes like hydro-sieves and centrifuges cannot separate fine particles in highly concentrated suspensions with a high selectivity. Disadvantages of these processes include the suspension concentration limit and high energy consumption.

In this work, the crossflow classification is investigated as a method to classify particles in a particle size range down to 1 μm. Crossflow classification utilizes the principles of the crossflow filtration. A suspension flows tangentially across the surface of a membrane. The pore size of the membrane is smaller than the particle size. Due to a higher pressure on the concentrate side, pure liquid permeates through the membrane (permeate flux). Meier et al. [1] suggest a new method of wet classification with the crossflow filtration. They recognized a selective deposition of fine particles. Altmann and Ripperger [2] developed a model for the particle deposition, which is based on a hydrodynamic model. The permeate flux creates a drag force on the suspended particles which transports them towards the membrane surface. The velocity gradient in the boundary layer of the crossflow creates a lift force on the particles, which transports them back into the bulk flow. In the present work, the model of the lift force by Rubin [3] is used. Both the drag and lift forces depend on the particle size. By regulating the crossflow rate and permeate flux, the separation size can be adjusted. The particles smaller than the separation size are deposited on the membrane, while the larger particles remain in the bulk flow. After rinsing of the system with a pure liquid, the particles deposited on the membrane can be resuspended by backwashing and discharged from the system. To find an optimal operating point, the influences of velocity and concentration were investigated.

2. Experimental setup

For the initial experiments, a lab-scale system. containing a pipe module with tubular membranes, which have a good resistance against backflushing, was constructed. The polypropylene (PP) membranes used have a nominal pore size of 0.2 μm. The effective membrane area was adjusted with different modules between 0.008 and 0.086 m^2. An automatic regulated pump produced a defined constant overflow velocity. By modulating the transmembrane pressure, the flux through the membrane was adjusted. To prevent decreasing flux caused by a particle layer on the surface, the transmembrane pressure was increased.

3. Experiments

A study of crossflow mean velocity on the separation size and selectivity was performed. The overflow velocity was varied between 0.1 and 2.5 m/s. The models for the lift force were adjusted for non-spherical particles and the influence of particle concentration and validated by experiments with different particle systems at multiple concentrations. The resuspension and discharge of the deposited fine particles was optimized. Factors affecting the stability of the resuspended particles after the classification such as particle charge were investigated.

4. Results

The experiments demonstrate successful classification of particles smaller than 5 μm. The influence of the pH on the stability of the suspension is shown. By adjusting the pH, resuspension of agglomerates into primary particles can be achieved.

[1] J. Meier, G.-M. Klein, Volker Kottke, Crossflow filtration as a new method of wet classification of ultrafine particles, Separation and Purification Technology 26 (2002) pp.43-50

[2] J. Altmann, S. Ripperger, Particle deposition and layer formation at the crossflow microfiltration, Journal of Membrane Sciense 124 (1997) 119-128

[3] G. Rubin, Widerstands- und Auftriebsbeiwerte von ruhenden, kugelförmigen Partikeln in stationären, wandnahen laminaren Grenzschichten, University of Karlsruhe, 1977

Copyright © American Institute of Chemical Engineers. All rights reserved.

514884 Study of Dry Beneficiation of Coal Particles in Air Dense Medium Fluidized Bed using CFD-DEM Method

Abstract:

Dry beneficiation of coal particles is an effective method of removing the ash, sulfide minerals and other harmful substances from raw coal. In this study, CFD-DEM method was adopted to study the dry beneficiation of coal particles in dense bubbling fluidized bed, where coarse-grained model is employed to simulate the motion of large amount of Geldart B magnetite powders and DEM is employed to model the motion of coal particles. In the CFD-DEM model, the weight function method and the center-count method are used to calculate the solid fraction and velocity of coal particles and coarse-grained magnetite powders, respectively. It was shown that (1) the simulated separation process of coal particles is in a good agreement with the experimental data under both of the superficial gas velocities of $1.5U_{mf}$ and $1.75U_{mf}$; (2) The light particles with densities of 1400kg/m^3 and 1700kg/m^3 are easily floated to the bed surface while the dense particles with densities of 2300kg/m^3 and 2500kg/m^3 settle down slowly to the bed bottom at the superficial gas velocity of $1.5U_{mf}$. But for the higher velocity of $1.75U_{mf}$, the particles with middle densities of 1700kg/m^3, 2000kg/m^3 and 2300kg/m^3 cannot be separated well due to the entrainment of large bubbles; (3) The force analysis indicates that the dense particles of 2300kg/m^3 and 2500kg/m^3 have a greater resistance force from dense medium particles in separation process. In summary, the CFD-DEM method is an effective method for studying dry beneficiation of coal particles.

Copyright © American Institute of Chemical Engineers. All rights reserved.

502842 Discovering the Fascinating World of Particle Technology

Abstract:

This talk will begin with a broad overview of the application areas of particle technology in the chemical process industry, and the role particle technologists play in addressing engineering problems. Specific examples of particle technology applications in sustainable energy generation and pharmaceuticals that involve nanoparticle manufacture, granular flow as well as fluidization and multiphase flow will be presented to highlight interesting and exciting aspects of this field. The intent is to assist the aspiring particle technologist in deciding â€œIs particle technology the right field for me?â€ The talk will then touch on opportunities in this education session for those interested in learning more about particle technology at all levels. The talk will then focus on the necessary techniques and skillsets a particle technologist uses in their practice, including theoretical, experimental and computational methods, and how best to deploy them in the field. It will conclude with a vision for the future of particle technology.

Copyright © American Institute of Chemical Engineers. All rights reserved.

508558 Chemical Looping for Reactive Separation

Abstract:

In this talk we will report on challenges and recent advances in calcium oxide based CO_2 sorbents (calcium looping). Alkaline earth metal oxides, such as CaO and MgO are an interesting class of CO_2 sorbents as their theoretical CO_2 uptake is very high. However, to obtain materials that show a high CO_2 uptake over many capture and regeneration cycles the structure and surface properties of the materials have to fulfill a number of criteria. As alkaline earth metal oxides capture CO_2 at elevated temperatures, these CO_2 sorbents can be integrated into catalytic processes. The so-called sorbent enhanced steam methane reforming will be presented as an example.

Copyright © American Institute of Chemical Engineers. All rights reserved.

508523 Demonstration of High Temperature and Pressure Gas-Solid Circulating Chemical Looping Reactor Systems for Syngas and Heat Generation â€“ Particle Reaction Analysis and Pilot Scale Test Results

Abstract:

Researchers are developing chemical looping technologies to convert carbonaceous fuels to high value chemicals and/or electricity with minimal CO_2 emission. These processes use a metal oxide or metal sulphate to partially or fully oxidize the fuel source to a desired product while being regenerated with air and/or steam in a separate reactor. The chemical looping redox reaction pathway is capable of high product yields without the need for molecular oxygen and minimizes gas product separation requirements. The Ohio State University (OSU) has advanced the chemical looping concept in the development of 2 pilot-scale demonstration plants for coal combustion with CO_2 capture and gaseous fuel conversion to high purity H_2 as well as multiple sub-pilot test units for syngas generation from biomass, natural gas, and coal. In each of these processes, a moving bed reducer reactor is used for partial or full conversion of the carbonaceous fuels to CO/H_2 and $CO2/H_2O$, respectively. A fluidized bed combustor reactor is used to regenerate the oxygen carriers to their original oxidation state. Interconnecting nonmechanical valves between reactors are used to control the solids circulation rate of the system for the mass balance of the redox reaction and heat balance for the system integration. The present paper discusses the unreacted shrinking core model (USCM) developed to simulate the reduction rate of the oxygen carrier particle which was used in the design of the moving bed reducers. This paper also summarizes the key developments of four chemical looping processes â€“ the $250kW_{th}$-$3MW_{th}$ syngas chemical looping (SCL) pilot plant, the coal direct chemical looping (CDCL) 25 kW_{th} sub-pilot unit and 250 kW_{th} pilot plant, and the 15 kW_{th} coal to syngas (CTS) and shale gas to syngas (STS) sub-pilot unit. Key process design features and experimental results from over 1,500 hours of cumulative operation will be discussed. The counter-current moving bed reducer reactor design the CDCL and SCL process ensure nearly full fuel conversion to CO_2 with minimal solid circulation and capability of producing high purity H_2. The co-current moving bed reducer reactor design in the CTS and STS processes provides a desirable gas-solid contacting pattern that minimizes carbon deposition and maximizes the syngas yield. The syngas produced by the CTS and STS processes can achieve adjustable H_2:CO ratio with optimized value of 2:1 with little CO_2, unconverted hydrocarbons, and steam, which is required for down-stream processes to produce liquid fuels and chemicals. During operation of the pilot plants, sliding mode control theory was adopted for modulating the aeration gas flow rate against the pressure fluctuations caused by the slugging bed combustor in order to maintain a constant solid circulation rate.

Copyright © American Institute of Chemical Engineers. All rights reserved.

509665 Design of Nano/Micro Structures of Hollow, Skeletal, and Porous Particles

Abstract:

Hollow silica nanoparticles exhibit unique properties such as thermal insulation, transparency, light diffusion, and so on which are derived from hollow interior surrounded by a silica shell. In order to improve these properties more, control of the silica shell structure in micro- and macro-level is important. The hollow nanoparticles are mainly prepared by template method, in which silica coating is performed on the solid or soft template surface and then the template is removed by chemical etching. Previously, gas adsorption analysis indicated that micro-pores (<2 nm) existed in the silica shell where molecules can pass through. Using the structure, selective deposition of tin oxide nanoparticles inside or outside the silica shell has been achieved. In addition, apparent density of the silica shell can be controlled between 1.45 and 2.20 g/cm^3 by adjustment of sol-gel conditions. The low apparent density of the silica shell significantly improved thermal insulation ability of the hollow nanoparticle /polyurethane film. While, macro-sized pores were also produced on the cubic shaped silica shell, by use of polarity difference in the reaction system. The skeletal silica nanoparticles have mostly hollow which divided by 12 silica nanoframes and are expected to protect organic substance (e.g., phosphor) from external stimuli. How to control micro- to macro-structures of hollow silica nanoparticles and how these functionalities have been achieved are discussed.

Copyright © American Institute of Chemical Engineers. All rights reserved.

508325 Unifying Principles of Product Design

Abstract:

Our mission is to address grand challenges in particle science and technology. These are related to the formation, processing and application of particulate systems along the five dimensions of size, shape, surface, structure and composition. Today, unit operations and product design are the two most important pillars of chemical engineering. Product design is the formation, formulation, handling, manufacturing, and characterization of complex multiphase products with specific properties. The applications define the required product properties which cover both classical fields of process technology in the chemical industry as well as new emerging fields of electronics, energy and environmental technologies, life sciences, materials science and engineering, nanotechnology, and photonic technologies highlighting the broad relevance of mesoscale science. The following five unifying principles of particulate product design are proposed which are widely applicable to many different kinds of products including solid, liquid, and even gaseous particles: i) Particle formation by top-down and bottom-up approaches, ii) interactions between the building blocks, iii) structure formation by self-organization and by transport processes, iv) characterization along the process chain, v) multiscale modelling and simulation. Results from the Erlangen Cluster of Excellence â€œEngineering of Advanced Materialsâ€ show that the joint venture of chemical engineering with materials science in concert with the basic sciences opens new prospects for all involved disciplines. In particular, chemical and biochemical engineering expands through particle technologies also in physics related fields of technology such as electronics, photonics, or 3D printing. Rigorous mathematical optimization methods based on predictive models for structure-property and process-structure functions are the key for the design of products, structures, and processes and will catalyze new possibilities for true design of particulate products.

References:

Peukert W., Segets D., Pflug L., Leugering G., Unifying principles of product design, Advances in Chemical Engineering 46 (2015), 1-81.

Copyright © American Institute of Chemical Engineers. All rights reserved.

507036 Improving Circulating Dehydrogenation Technology through Optimization of Fluidization

Abstract:

The dehydrogenation of light alkanes to corresponding light olefins is industrially an important process. One of the commercial alkane dehydrogenation circulating fluidized bed reactors was underperforming for quite some time. In addition, the catalyst losses were significantly higher than the design value. Several aspects related to fluidization, catalysis and process engineering were investigated to understand and resolve these issues in the Plant. The co-authors of this study focused on the fluidization aspects of the process, and benchmarked equipment design as well as operating conditions against industrial best practices. The team extensively studied the design of feed distributors, cyclones, internals, strippers, etc. Furthermore, attrition properties were also assessed for different types of catalyst used in the unit. Several recommendations and improvement modifications were proposed by the team in order to increase olefin production and to reduce catalyst losses significantly. This study provides insights into the improvement of fluidization behavior and systematic approach followed by the team to determine underlying root causes in the circulating fluidized bed unit.

Copyright © American Institute of Chemical Engineers. All rights reserved.

508509 Hydrodynamic Characteristics of High Pressure High Temperature Counter-Current Moving Bed Chemical Looping System Circulated with Geldart Group D Oxygen Carrier Particles

Abstract:

The chemical looping combustion processes developed at the Ohio State University (OSU), which uses Fe-based oxygen carrier and a counter-current moving bed reducer design, have the capability of achieving nearly 100% carbon capture and the flexibility of co-generating high purity hydrogen and electricity. To accommodate the moving bed operation, the oxygen carriers are classified as Geldart Group D particles with an average particle diameter of 1.5 mm and a density of 2500 kg/m^3. The system consists of a reducer operated as counter-current moving bed reactor, and a combustor operated as a fluidized bed reactor, to perform reductionâ€“oxidation reaction cycles for converting fuels carbon dioxide (CO_2). An oxidizer can also be integrated into the system, between the reducer and combustor reactors, to produce a nearly pure stream of hydrogen (H_2). The oxidizer is operated in the counter-current moving bed mode similar to the reducer reactor. An L-valve is used in the chemical looping systems to transport the solids between the moving bed reactors and the fluidized bed combustor where the aeration gas flow rate is directly correlated to the global solid circulation rate. An iso-kinetic device is installed on the standpipe of the moving bed side to measure the solids circulation rate of the system. Thousands of hours of operations with different solids circulation rates and gas processing capacities have been performed on sub-pilot and pilot scales of the systems under high operating pressures (>10 atm) and temperatures (>1,000 °C). The hydrodynamic characteristics of oxygen carrier particles operated in the systems, such as pressure drops and fluidization regimes, under different operational conditions are obtained. The operational ranges of gas flow rates for the reducer and the combustor under different pressures and temperatures are obtained. The characteristic relationship between solids circulation rate and aeration gas of L-valve under different temperatures is investigated. The gas demands for fluidizing the combustor and transporting solids through the riser under varing operating temperatures and pressures are also investigated. Fluidization regimes of the combustors are identified. The slugging behavior of the combustor with high height-to-diameter ratio under elevated temperature conditions is also investigated with the aid of ECVT measurement.

Copyright © American Institute of Chemical Engineers. All rights reserved.

509052 Defluidization Behaviour of Industrial Reactive Powders at High Temperature

Abstract:

Fluidized bed reactors are used in a wide range of industrial chemical processes, and they are particularly useful in high temperature systems thanks to their ability to provide high heat transfer rates and rapid solids mixing which lead to isothermal and more controllable conditions. Their applications spread the environmental, chemical, energy and process industries. For example, they are a key technology in the petroleum industry, catalytic chemicals synthesis processes, combustion and gasification of solid fuels (coal, wastes and biomass), metals productions and many more.

Operation below incipient fluidising conditions in combustion â€" type operations can lead to particle agglomeration and sintering which change the characteristics of the bed, leading to possible bed defluidisation unscheduled process downtime and additional costs. However, the details of the physical processes under such conditions are still not well understood. Despite the countless researches that have been done so far, such processes still disclose unrevealed features and challenge the researchers worldwide.

The objective of this work is to assess the effect of high temperature on the flow properties and fluidization behaviour of such systems using carbon coke and two different titanium ores. To this end, a simple approach for the early detection of agglomeration and defluidisation in the regime below incipient fluidization - when the carbon combustion initiates the defluidization- will be followed. Such an approach is based on the simultaneous measurements of local temperatures and the pressure drop across the particle bed.

Copyright © American Institute of Chemical Engineers. All rights reserved.

508544 Fluidization of Cohesive Materials Using Modified Solids Mechanics Models

Abstract:

Fluidization of cohesive materials is difficult to achieve. These materials tend to form channels during fluidization. They also form structures within the fluid bed that lift surround bulk material and can develop pressures and pressure gradients that are larger than the hydrostatic gradient normally experienced in fluid beds. We want to approach this phenomenon from a solid mechanics point of view. One can model the channel formed in a fluid bed as a pipe or a rathole in material that has strength and treat it as a slope stability problem â€" except that the bulk material has pressure gradients that modify the stability of the channel surface. The basis of these equations is part of a previous work. The size of stable ratholes with gas pressure gradients can be computed using this past work. However, the motion of material in the channel during fluidization can induce a shear that is not compatible with the traditional assumptions for calculation of rathole stability from a solids mechanics point of view. Traditional theory assumes an unconfined state at the surface of the rathole and implies unique conditions on principle stress at this location. This assumption is not valid when a shear stress exists on this surface. We have reformulated the rathole stability equations with the assumption of a local shear stress at this boundary and also included gas pressure gradients. The resulting equation was designed to predict the stability of a channel formed in a bulk solid subject to movement in the channel and local gas pressure gradients. This is closer to the conditions that one would expect in the fluidization of a cohesive material.

We constructed a fluid bed and measured the standard fluidization curves as well as the size of channels formed in the bed. We also measured the unconfined yield strength of the material at low solid contact pressures and then compared this data to the values predicted from the modified rathole theory, including movement of material. An FEM analysis of gas pressures and solids stresses surrounding the channel was coupled with the results of the rathole equations to predict the channel size. We also measured the settlement time of cohesive materials as a function of the unconfined yield strength, and then created a settlement model that included gas loss through channels that formed. Finally, we correlated this behavior with the expected aerated rathole stability predictions. The results are presented in this paper.

Copyright © American Institute of Chemical Engineers. All rights reserved.

508861 Dynamic Simulation of Interconnected Solids Processes

Abstract:

1. Introduction

A novel flowsheet simulation system, aimed at the dynamic modelling of complex production processes mainly in the area of solids processing, is being developed within the Priority Program SPP 1679 "Dynamic simulation of interconnected solids processes (DYNSIM-FP)" of the German Research Foundation (DFG) since July 2013 [1]. The Priority Program consists of 29 separate projects from different German universities and is thematically divided into four main groups:

- A: Development and implementation of new dynamic models of individual apparatuses and process substeps;

- B: Calculation of material parameters and development of models to describe properties of solids materials;

- C: Development of advanced algorithms and methods for dynamic simulation of complex process structures;

- Z: Development of a novel framework for the flowsheet simulation of dynamic processes.

This flowsheet simulation framework, named Dyssol (acronym for the DYnamic Simulation of SOLids processes), is presented in this contribution. The main objective of the system is to combine models, methods and algorithms, developed within the whole SPP 1679 program, into a single modelling framework that allows performing dynamic modelling of complex process structures involving solid, liquid and gaseous phases as well as their mixtures.

2. Calculation algorithm and applied methods

The Dyssol simulation system implements a sequential-modular approach [2] (SMA), where each unit is relatively independent from others and can be computed separately by its own numerical method or calculation algorithm using an individual simulation time step. Obtained results after computation of each unit are transferred between units according to the flowsheet structure. This approach leads to a relative independence between models and provides high flexibility of the simulation system itself.

The SMA implies sequential computation of models in a certain order, so it is necessary to know all inlets of a model to start calculation. Flowsheets, which consist solely of serially connected modules, satisfy this condition by default. On the other hand, applying of SMA is complicated if there are recycle streams in a process structure. In this case a flowsheet must be additionally converted into sequential form by tearing recycle streams [3]. Tearing here means search for recycle streams and setting some initial values into them.

For the dynamic simulation a modified waveform relaxation method (WRM) [4] is used. The whole simulation time is divided into smaller non-constant time intervals, on which convergence can be reached faster. Units are solved separately on each time window using some initial guess for the solution. Calculations are repeated on the current time interval until the convergence is reached. Then, the system proceeds to the calculations on the next time window. To initialize each next time window extrapolated values from the previous time intervals are used. The number of WRM iterations, needed to reach convergence, depends on the difference between the extrapolated and actual results. A set of extrapolation algorithms has been implemented in Dyssol to initialize the first iteration on each time interval.

3. Parallelization methods

Dynamic flowsheet simulation is a computationally demanding process. Sequential-modular algorithm of the Dyssol makes it relatively easy to implement different parallelization mechanisms [4] that can significantly reduce calculation time. Several strategies are implemented in the simulation system Dyssol:

- Parallelization of basic functions of the simulation system, such as matrix transformation and mixing of material streams.

- Parallel calculation of the model itself. Feasibility of this approach depends on the particular model, and the specific implementation is the task of a model's developer.

- Simultaneous calculation of individual time points for steady-state models. If the outlet parameters of model are independent of its internal state, and are determined only by the input parameters, model state at different time points can be computed in parallel.

- Parallel calculation of models, which are located on independent branches of a flowsheet and thereby are not directly dependent from each other.

4. Handling of multidimensional distributed parameters

One of the main challenges during the simulation of solids processes is related to the dispersity of granular materials: the solid phase can be distributed along several interdependent properties, such as size, shape, moisture content, density, etc. forming a multidimensional set of distributed parameters. On the other hand, each operation on the material flow, such as separation, mixing, agglomeration, etc. may require a change in one or more distributed parameters. In this case it may be necessary to calculate all dimensions from a multidimensional set even if the value of a single parameter had been changed. Therefore, it should be considered that during the simulation interdependency between all parameters should not be lost. This means that applying the same methods for calculation of solids as for liquid-gaseous systems is impossible or can lead to their incorrect handling.

All parameters of the granular materials are represented in the simulation system Dyssol in the discretized form. The entire interval, on which the parameter is determined, is represented as a set of shorter intervals. To describe the distribution each such interval is associated with the number or mass of particles whose parameters fall within this interval. The discretization scheme must be fine enough that the numerical errors were small compared to the accuracy of the used models.

To represent interdependency between several distributed parameters multidimensional matrixes [4, 5] are used, where the number of dimensions is equal to the number of parameters. Each cell of such matrix contains the mass fraction of the solid material with certain combination of parameters, whereas the total sum of all entities of the matrix is equal to one.

For storing of such discretized multidimensional distributed parameters a sparse data format is used. The multidimensional matrix is represented by a flat tree data structure, where each level of this tree is one of the dimensions. If there are some classes, which are empty (contain zeroes), there is no need to store data on lower levels. This approach provides an opportunity to reduce memory consumption and calculation time. For more efficient usage of this structure dimensions are arranged in such a way, that the more zeroes a dimension contains, the higher it is placed in the structure.

For correct handling of solids an approach with transformation matrices is used [4, 5]. Each cell of a transformation matrix describes the mass fraction of material that passes from one class with a particular combination of parameters to some other class with another combination of parameters. This concept allows preserving information about all parameters, even those which are not considered or not changed in the current model. Thus, instead of an explicit calculation of the model's parameters transformation matrix can be generated and applied to obtain output stream in each unit.

When using this approach, the model of each particular unit is slightly complicated because not the final distribution of parameters is being calculated, but the laws of their transformation. However, this concept ensures that all information about the distributed parameters of granular materials will be retained during the computation and properly considered.

Acknowledgement

We gratefully acknowledge the financial support of the German Research Foundation (DFG) within the priority program SPP 1679 “Dynamic simulation of interconnected solids processes DYNSIM-FP”.

References

[1] www.dynsim-fp.de (2017).

[2] Hillestad, M., Hertzberg, T.: Convergence and stability of the sequential modular approach to dynamic process simulation. Comput. chem. Engng. 12, 1988, 407-414.

[3] Towler, G., Sinnott, R.: Chemical engineering design: principles, practice, and economics of plant and process design. Second edition. Butterworth-Heinemann Press, 2013.

[4] Dosta, M.: Hamburg-Harburg, Techn. Univ., Diss., 2012.

[5] Pogodda, M.: Hamburg-Harburg, Techn. Univ., Diss., 2007.

Copyright © American Institute of Chemical Engineers. All rights reserved.

509589 Simulation of Powder Segregation Phenomena during Filling and Emptying of an Industrial Storage Hopper

Abstract:

This presentation describes a numerical simulation of segregation of a powder mixture when a storage bin is used in a powder process. A continuum framework tracks composition changes inside and exiting the bin during its dynamic operation, including filling and emptying. The simulation treats particle segregation on moving free surfaces and includes velocity profiles within the powder during discharge. The simulation combines historical treatments of powder flow (mass and funnel flow) from the 1960â€™s with the recent developments in powder segregation modeling from Northwestern University. This simulation was programmed using gPROMS and therefore it can be connected to other unit operations within the gSOLIDS library for an overall dynamic powder process simulation. We will show how this modeling approach is useful to industry in mitigating segregation risk with examples.

Copyright © American Institute of Chemical Engineers. All rights reserved.

506839 Atomic Layer Deposition for the Synthesis of Nanostructured Catalysts

Abstract:

Atomic layer deposition (ALD) is a thin film growth technique that uses cycles of alternating, self-limiting chemical reactions between gaseous precursors and a solid surface to deposit material in an atomic layer-by-layer fashion. Saturation of the individual ALD surface reactions and gaseous diffusion of the chemical precursors ensure that all exposed surfaces of a template become uniformly coated. These attributes, coupled with a broad palate of available materials, make ALD a versatile technique for synthesizing catalysts. Conceptually, ALD can be used to build a catalyst as a series of discrete layers which perform specific functions such as tuning the porosity, serving as the catalyst support, providing catalytic activity, and imparting thermal stability. This capability allows the physical structure and porosity of a template to be decoupled from the surface chemistry and catalytic properties. Consequently, the same catalyst can be synthesized simultaneously on a variety of templates such as a planar single crystal to facilitate detailed characterization, or a high surface area powder to boost conversion for catalytic testing.

In addition to depositing continuous layers, ALD also enables the synthesis of nanoparticles. Using ALD, the nanoparticle size can be modulated by adjusting the number of cycles performed, and the number of nanoparticles can be controlled by pretreating the surface to change the concentration of surface functional groups on which the ALD initiates. By alternating between different ALD surface chemistries, mixed-metal nanoparticles (e.g. Pt-Pd) can be formed. Furthermore, the layer-by-layer nature of ALD enables specific nanostructures for these mixed-metal nanoparticles such as core/shell (Pd/Pt or Pt/Pd) and well-mixed alloy. Finally, ultrathin ALD metal oxide layers applied after the metal nanoparticle ALD can stabilize the nanoparticles to inhibit sintering at elevated temperatures. Quite remarkably, these metal oxide â€œovercoatsâ€ grow selectively on specific sites on the noble metals and on the support between the nanoparticles such that the catalytic activity is maintained or even enhanced.

Copyright © American Institute of Chemical Engineers. All rights reserved.

509584 The R & D on Engineered Particles for Functional Materials in China

Abstract:

This is an overview on the R & D activities of functional powder processing from the particle level. China needs this kind of technologies to support its new material industry and improving the qualities of commonly used materials.

It starts from the national policy in this 10-year action plan designed to transform the structure of processing industry and increase the quality of â€œMade in Chinaâ€. The Key Developing Sectors are: Aeronautics/Astronautics, New Generation IT, Machine Tools and Robots, Marine Engineering, Advanced Rail Transportation, New Energy Vehicles, New Energy, and New Materials.

Particle coating and hybridization is an important approach to make functional powders. Newly developed equipment is introduced. Copper/alumina, PE/CNT composite particles are prepared. Particle shape modification is another technology to increase the powder flow-ability and higher tap-density. The technology is introduced with the analysis of the process on graphite powder process for Lithium-ion Battery Anode.

Copyright © American Institute of Chemical Engineers. All rights reserved.

509079 Creating a Pathway to Sustainability: The Critical Contribution of Particle Technology Research

Abstract:

Maintaining adequate future supplies of energy and water requires the creation of a pathway to sustainable energy and water that is both multi-faceted and complex. An effective pathway aims to improve the reliability, security, and affordability of energy and water by: utilizing a least-cost strategy to reduce the negative impact of energy consumption on climate change and water availability; emphasizing the gradual decarbonization of the global energy system; increasing energy efficiency and conservation; and preserving natural resources and the environment.

A successful pathway to sustainability must include short-term, transitional, and long-term plans that must be launched simultaneously. This presentation will discuss the critical contribution of particle technology research to the successful implementation of these plans.

The role of circulating fluidized bed (CFB) research in carbon capture processes and particle pulverization as a recycling tool for polymeric and elastomeric materials as examples of the short-term plan will be presented.

As an example of the transitional plan, the impact of fluid-particle flow and computational fluid dynamics (CFD) research in the production of natural gas from unconsolidated hydrate reservoirs will be discussed.

The contribution of particle technology to the long-term plan, including the effect of rain droplets on wind turbine performance and the role of the fluid-particle mixture as a very effective heat transfer medium for concentrated solar power (CSP) systems will be presented.

Copyright © American Institute of Chemical Engineers. All rights reserved.

507204 Chemical Looping Combustion, Gasification and Reforming â€“ Particle Technology Perspectives

Abstract:

The concept of chemical looping reactions has been widely applied in chemical industries. Fundamental research on chemical looping reactions has also been applied to energy systems. Fossil fuel chemical looping applications were used with the steam-iron process for coal from the late 1800s to early1900s and were demonstrated at a pilot scale for synthetic natural gas production with the IGT HYGAS Process and the CO_2 Acceptor Process in the 1960s and 1970s. There are presently no chemical looping processes using carbonaceous fuels in commercial operation. Key technical factors that determine commercial viability of the technology lie in the sustainability of the reactivity and recyclability of the metal oxide oxygen carriers and the ability of configuring the reactor assembly for optimal operation and control of the solids flow system. The chemical looping system is developed on the circulating fluidized bed platform. The successful deployment of this technology requires thorough knowledge of two interconnected fields, i.e., metal oxide reaction engineering and particle science and technology. With now CO_2 emission control of great concern and process conversion efficiency enhancement of great interest, activities on research and development of chemical looping technology have resurfaced.

Specifically, chemical looping technology is a manifestation of the interplay among such key elements of metal oxide reaction engineering and particle science and technology as particle synthesis, reactivity and mechanical properties, flow stability and contact mechanics, gas-solid reaction engineering and particulates system engineering. This presentation will describe the fundamental and applied features of modern chemical looping technology in the context of the circulating fluidized bed platform that utilizes fossil and other carbonaceous feedstock. It will discuss particle reaction engineering represented by reaction chemistry, ionic diffusion mechanisms, metal oxide synthesis and thermodynamics, reactor design, and system engineering along with energy conversion efficiency and economics of the chemical looping processes for full and partial or selective oxidation for electricity, fuel and chemicals production. The Ohio States University has developed a number of advanced chemical looping combustion, gasification and reforming processes which will be highlighted in this presentation. Potential for solar based chemical looping technology will also be discussed.

Copyright © American Institute of Chemical Engineers. All rights reserved.

507396 Riser Dynamics - a Comparison of Scale

Abstract:

Risers in gas-solids circulating systems are used extensively in the petrochemical and energy industries as well as being proposed for advanced technologies from chemical looping combustion/gasification. The performance of these units has been linked to the meso-scale and the dynamic behavior of the flow field. New experimental data has been obtained in a 0.1 m diameter riser with a 5.3 m height. The particle is an 850 µm high density polyethylene with a density of 860 kg/m^2. Experiments were conducted primarily in the core-annular regime. In addition to standard differential pressure measurements along the riser, three-dimensional measurements of the solids volume fraction were obtained using a Tech4Image Electrical Capacitance Volume Tomography (ECVT) system. The dynamic structure of the riser solids fraction is obtained by analyzing the time series data for each voxel in the flow field. Specifically, the data were analyzed using higher level statistics (skewness and kurtosis) as well as chaotic correlation dimension and the entropy. These values are compared with published ECVT data using the same material from a 0.3 m diameter riser with a 16 m height. The effect of system scale is analyzed and discussed.

Copyright © American Institute of Chemical Engineers. All rights reserved.

507361 Detailed Numerical Simulation of a Full-Loop Circulating Fluidized Bed Under Different Operating Conditions

Abstract:

Detailed numerical simulation of a full-loop circulating fluidized bed under different operating conditions

Yupeng Xu[1,*], Jordan Musser[1], Tingwen Li[1,2], William A. Rogers[1]

1 National Energy Technology Laboratory, Morgantown, WV 26505, USA

2 AECOM, Morgantown, WV 26505, USA

Abstract

Circulating fluidized beds (CFB) are widely employed in industry for a wide variety of gas-solid contacting operations. Over the past few decades, CFB hydrodynamics have been studied using various experimental techniques and by numerical simulations. Recently, several efforts have modeled full-loop CFBs using the two-fluid model (TFM) whereby gas and solids are treated as interpenetrating continua. However, this technique only resolves solid motion at the computational cell level. With the advancement of high-performance computing, higher fidelity models, like Computational Fluid Dynamics-Discrete Element Method (CFD-DEM) which tracks the trajectory of individual particles, have become increasingly popular as research tools. This is because particle-scale information including residence time, collision forces, and dispersion intensities are readily available for detailed analysis of complex flow phenomena.

In this first of its kind work, both experimental and computational studies of a benchtop, full-loop circulating fluidized bed are conducted. Experimental data include pressure drops across different components, solids circulation rate measured using Particle Image Velocimetry (PIV), and standpipe inventory height. Numerical simulations were performed with MFIX-DEM, a coupled Computational Fluid Dynamics-Discrete Element Method (CFD-DEM) solver. Comparison of experimental and simulation data shows good agreement with respect to system component pressure drops and standpipe inventory height. Additionally, the effects of drag model selection and particle representation (mono-sized particles versus a particle size distribution) are also reported.

Figure 1. Experimental setup and operation conditions (SLPM)

Figure 2. Direct comparison between experiments and simulations under different operating conditions (Left: experiments, right: simulation)

Copyright © American Institute of Chemical Engineers. All rights reserved.

508046 Generating Macroscopic Quantities of Particle-Fluid Flows of Coarse Particles By an Averaging Method: Selection of Sample Sizes and Test of Solid Pressure Correlation

Abstract:

Generating macroscopic quantities of particle-fluid flows of coarse particles by an averaging method: Selection of sample sizes and test of solid pressure correlation

Q.F. Hou[1], Z. Y. Zhou[1], J. S. Curtis[2], and A.B. Yu[1,3]

[1]ARC Research Hub for Computational Particle Technology, Department of Chemical Engineering, Monash University, Clayton, VIC 3800, Australia

[2]College of Engineering, University of California, Davis, One Shields Ave, Davis, CA 95616, USA

[3]Centre for Simulation and Modelling of Particulate Systems, Southeast University - Monash University Joint Research Institute, Suzhou 215123, PR China

Abstract

Macroscopic quantities of particle-fluid flow such as solid density and stress tensor can be obtained from microscopic variables such as particle positions and interaction forces through averaging methods. In using average methods, spatial and temporal sample sizes play an important role. However, how to properly select sample sizes is not clearly understood. Furthermore, the validity of the obtained correlations using the obtained macroscopic quantities has not been well examined. By using the combined approach of computational fluid dynamics and discrete element method, this work is attempted to give answers to these two concerns. First, the effect of spatial and temporal sample sizes on averaged macroscopic quantities is discussed. Guidance for the selection of sample sizes is provided to generate macroscopic quantities which are compatible with continuum mechanics. Then the correlation between solid pressure and solid fraction is established and compared to the correlations available in the literature. It agrees quantitatively with experimental observations and semi-empirical correlations but agrees only qualitatively with kinetic theory predictions. The evidence proves that the averaging method is an effective approach in linking microscopic and macroscopic properties of particle-fluid flows. If the sample sizes are selected properly valid correlations can be formulated for continuum modeling.

Key words: fluidized beds, granular media, particle-fluid flows

Copyright © American Institute of Chemical Engineers. All rights reserved.

508263 Structured Flow in Gas-Solid Fluidized Beds: Particle Clustering and Bubble Self-Organisation

Abstract:

Fluidized beds are widely used to put in contact solid and gas phases in a variety of industrial processes. The interaction force between the gas and the suspended solids and the energy dissipated in particle collisions originate complex flow features that are dependent on the solids properties e.g. spouting, channelling, jetting or bubbling. Despite the advance in design and control, multi-phase flow models still struggle to describe many of these structures and scale up requires from a significant amount of historical practice and experimentation. The performance of a bubbling fluidized bed for instance is heavily reliant on a good level of mixing, associated to the observed solid circulation and the bubble characteristics e.g. size, velocity. From a fundamental perspective, the formation of bubbles can be explained by the instability of the coupled momentum equations or any similar phenomenological model used to study the bed response and the pressure fluctuations associated to nucleation, coalescence and rupture. While this type of analysis allows the formulation of a theoretical framework [1], important effects such as particle polydispersity, segregation, particle inertia and shape remain poorly understood. Furthermore, hydrodynamic models leave open questions regarding the behaviour of dense granular systems where multiple particle contacts and local anisotropy become important in the transport of momentum [2]. This constitutes the general pitfall of numerical frameworks based in the kinetic theory of granular flow. While they remain the only alternative to support the design of full scale units, most kinetic models rely on a microscale description of particle interactions that is inconsistent with the complex granular rheology of granular media in the dense and quasi-static regimes [3].

Allowing for more degrees of freedom at the design stage is a promising way to introduce a higher level of control in the operation of bubbling fluidized beds. In recent years, several alternatives have been put forward to manipulate the time and spatial scales of particle and particle-fluid interactions at a micro- and mesoscale in order to render a more reproducible flow structure in the macroscopic system. In addition to other methods based in modifying the spatial distribution of forces in the bed e.g. particle-particle applying electric fields or particle-fluid using fractal injectors [4], the pulsation of the inlet gas has been identified as a promising candidate to structure the flow in bubbling beds. Pulsating the gas flow introduces a temporal dynamic in the drag force that can supress the natural instability of the gas-solid flow system. Ripples in sandy beaches or dunes in deserts are good examples of dynamic granular structures that emerge as a response to a periodic perturbation in the flow around them, e.g., tides and waves, wind gusts and eddies. Transferring the same principle into the operation of quasi-2D fluidised bed, we have demonstrated that under certain conditions it is possible for gas bubbles to reorganise into a regular pattern that propagates throughout the bed into an stable large scale flow structure [5]. Under these operating conditions one has a much tighter control over the resulting bubble size, velocity and separation, thus opening the way for an entirely new concept in unit control and design.

This contribution investigates how dynamic bubble flows emerge in the fluidization of Geldart B particles under a pulsed airflow. At certain conditions, bubbles rearrange into a triangular tessellation with a narrow bubble size and wavelength distribution, independent of the bed width. We discuss the role that the dissipation of energy through friction plays in the bubble nucleation mechanism and how the local changes in rheology can stabilise the bubble pattern. We report a series of experimental data in quasi-2D pulsating beds along numerical studies under different airflow conditions, inter-particle friction factors and bed dimensions. The transition from the formation of a surface waves in a shallow layer to the nucleation of bubbles in thicker beds is related to the analysis of large clustered regions appearing in the wake and around pairs of rising bubbles. A comparison between continuous and discrete numerical formulations given by Eulerian-Eulerian (TFM) and Eulerian-Lagrangian (CFD-DEM) frames is used to showcase the limitations of classical kinetic and rheology models in predicting the behaviour of the granular flow in a dense and plastic regime. The inherent difficulty of a kinetic model to deal with correlated particle velocities and the resulting anisotropic distribution of contacts can explain why the common closures for the solid stress fail to capture how bubble self-organise under dense conditions[5]. The explicit solution of the rheology through an Eulerian-Lagrangian frame however succeeds in tracking the local changes in granular rheology and is shown to reproduce the observed bubble dynamics [6].

References

[1] Batchelor G.K. (1993). Secondary instability of a gas-fluidized bed, J. Fluid Mech. 251, 359-311.

[2] Chialvo S., Sun J., Sundaresan S. (2012). Bridging the rheology of granular flows in three regimes Physical Review E 85, 021305 1-8.

[3] Chialvo S., Sundaresan S. (2013) A modified kinetic theory for frictional granular flows in dense and dilute regimes. Physics of Fluids 25, 070603.

[4] Coppens M-O, Ommen J.R, (2003). Structuring chaotic fluidized beds, Chem. Eng. J. 96 117-124.

[5] Wu K, De Martin L., Mazzei L., Coppens M-O. (2016). Pattern formation in fluidized beds as a tool for model validation: A two-fluid model based study. Powder Tech., 295, 35-42

[6] Wu K, De Martin L., Coppens M-O. Pattern formation in pulsed gas-solid fluidized beds â€" The role of granular solid mechanics. Chemical Engineering Journal (2017). In Press. https://doi.org/10.1016/j.cej.2017.05.152

Copyright © American Institute of Chemical Engineers. All rights reserved.

509348 Studies on the Mechanism and Regulation-Control Method of a High Density Gas-Solids Circulating Fluidized Bed Downer

Abstract:

The efficient conversion and optimal utilization of heavy oil has great practical significance to the development of national economy. Fluid catalytic cracking (FCC) is a critical approach to deal with heavy oil. Riser, as the key part of the FCC unit, is characterized by the non-uniform â€œcore-annualsâ€ flow structure with relatively significant gas and solids backmixing which may result in low reactant conversion and product selectivity. Downer reactor, in which gas and solids move downward co-currently, has unique features such as the plug-flow reactor performance and relatively uniform gas-solids distribution compared to the riser reactor. Downer is therefore acknowledged as a novel multiphase flow reactor with great potential in dealing with heavy oil. However, low solids holdup and poor initial gas-solids contacting hinder the promotion and application of downer reactor. Mechanism of increasing solids holdup and approaching high density downer have not been well understood until now. This study starts from the entrance of the downer to in-deep investigate its effects on high density operating condition and to study the contributions of the downer entrance to accomplishing high solids holdup and good gas-solids contacting and in downer reactor. On the other hand, all parts of the entire circulation system have been optimized in order to develop a regulation and control method to obtain and maintain a stable operating of the high density downer. Based on the above work, the mechanism of achieving the high density downer has been demonstrated, a multi-scale computational model has also been established to preciously describe the hydrodynamic characteristics and the scale-up of high density downer has been illustrated all of which can provide the fundamental data and basic theoretical support to develop a technology of downer catalytic cracking of heavy oils and extend the applications of high density downer reactors.

Copyright © American Institute of Chemical Engineers. All rights reserved.

508024 Turbulent Closure Models for Multiphase Fluids

Abstract:

Turbulent multiphase flows are encountered ubiquitously in the process industry. Examples include flows within hydrocyclones and fluidized bed reactors. The potential for simulating these flows has primarily occured because of advancements in computational hardwarre and software. However, current deficiencies in the turbulences models for mixing seriously limit the practical utility of computational methods as an enabling technology. The proposed presentation will extend a recently developed realizable closure for the Reynolds stress of a single phase fluid to a mixture of immiscible fluids. Most significantly, the theory guarantees realizable behavior of the multiphase Reynolds stress for all turbulent flows in inertial frames and in non-inertial frames.

Copyright © American Institute of Chemical Engineers. All rights reserved.

A CFD-DEM Study of Bubble Properties in Fluidzed Beds Operated with a Single Jet

Siddhartha Shrestha, Jieqing Gan, Aibing Yu and Zongyan Zhou*

Laboratory for Simulation and Modelling of Particulate Systems, Department of Chemical Engineering, Monash University, Victoria 3800, Australia

Abstract

Bubbles properties play an important role in determining the characteristics of gas-solid fluidized beds and thus influence the bed performance significantly. In this work, the bubble properties are studied by the combined approach of CFD and DEM for bubbling gas-solid fluidized beds operated with a continuous central jet. The process of the continuous injection of a central air jet to the bed is successfully reproduced featured with the formation of series of bubbles which rise through the bed and burst at the bed top. The bubbles are categorised as two types: primary/leading bubbles and trailing bubbles. The size of the primary bubble keeps on increasing with bed height while the size of the trailing bubble decreases since trailing bubble is coalescing with the primary bubble. The primary bubbles although formed at different times are of similar shape while the trailing bubbles are generally elongated in shape. The bubble velocity is distinctly different for primary and trailing bubbles.

Keywords: bubble formation, bubble size, bubble shape, single jet, CFD-DEM

1. Introduction

Fluidization technology has been widely used in many industrial applications such as food, pharmaceuticals, energy and environment since they are associated with good heat and mass transfer. The flow behaviour in gas-solid fluidized beds are very complex because of the fluid-particle and particle-particle interactions. Bubbles often exist in fluidized beds and are termed as "motor of fluidization"[1]. They govern hydrodynamics of fluidized beds, induce gas-solid mixing and promote heat and mass transfer [2]. Thus, a good understanding of bubble formation at an orifice and its dynamics is essential to aid in design and development of gas-solid fluidized beds. Various experimental studies have been carried out to portray bubble formation at an orifice [3] and to establish theoretical models [4, 5]. Besides, some experimental studies also used non-invasive techniques to study time evolution of bubble characteristics [3], the pressure wave propagation and attenuation behaviour [6] and bubble induced particle mixing [7]. In recent years numerical techniques have been widely used to study gas-solid fluidized beds and have become a significant approach to characterize bubble behaviour. Numerical techniques for gas-solid systems can be either continuum- or discrete-based with respect to the solid phase. The former is called two-fluid model (TFM), where both gas and solid phases are considered continuous and fully interpenetrating. The latter is referred as discrete element model (DEM) where particle phase is modelled as discrete phase and gas phase as continuous. Several authors have used TFM to study bubble formation at a single orifice and its dynamics [8-12]. However, DEM simulations have advantages over TFM simulations because the solid phase is modelled much more accurately, with respect to friction [13-15]. DEM studies in single jet fluidized beds were carried out to show that the size of a bubble is sensitive to cohesive inter-particle force [15], gas injection velocities, particle properties (size, shape and density) and bed dimensions [14-17], to depict the bubble induced particle mixing [18] and to formulate model for bubble-to emulsion phase mass transfer [19].

* *Corresponding author: zongyan.zhou@monash.edu; Tel: +61 3 9905 0846*

However, only the properties of the single bubble have been studied previously via CFD-DEM. The continuous injection of a central air jet to the bed leads to formation of series of bubbles which rise through the bed and burst at the bed top. In this process different types of bubbles are generated, and their behaviour are significantly different. To date, how the bubble properties vary in the bed when a continuous jet of gas is injected from a single orifice has not been fully investigated and the understanding is still lacking. Therefore, the objective of the present work is to study the bubble properties (size, shape and velocity) and force distribution in gas-solid fluidized beds operated with a continuous central gas jet by using CFD-DEM.

2. Models and simulation conditions

In CFD-DEM, the solid phase is based on the DEM and gas phase is modelled as CFD. The governing equations for particle i with mass m_i and moment of inertia I_i can be written as

$$m_i \frac{dv_i}{dt} = \mathbf{f}_{pf,i} + \sum_{j=1}^{k_i} \left(\mathbf{f}_{c,ij} + \mathbf{f}_{d,ij}\right) + m_i\mathbf{g} \tag{1}$$

$$I_i \frac{d\omega_i}{dt} = \sum_{j=1}^{k_i} (\mathbf{M}_{t,ij} + \mathbf{M}_{r,ij}) \tag{2}$$

where $\mathbf{v}_i$ and $\boldsymbol{\omega}_i$ are the translational and angular velocities of the particle, respectively, and k_i is the number of particles interacting with the particle. The forces involved are: particle fluid interaction force $\mathbf{f}_{pf,i}$, the gravitational force $\mathbf{m}_i\mathbf{g}$, and the inter-particle forces between the particles, which include elastic force $\mathbf{f}_{c,ij}$ and viscous damping force $\mathbf{f}_{d,ij}$. The torques acting on particle i by particle j include: $\mathbf{M}_{t,ij}$ generated by the tangential force and $\mathbf{M}_{r,ij}$ commonly known as the rolling friction torque.

The governing equations for gas phases are given as,

$$\frac{\partial \varepsilon_f}{\partial t} + \nabla \cdot (\varepsilon_f \mathbf{u}) = 0 \tag{3}$$

$$\frac{\partial(\rho_f \varepsilon_f \mathbf{u})}{\partial t} + \nabla \cdot \left(\rho_f \varepsilon_f \mathbf{u}\mathbf{u}\right) = -\nabla p - \mathbf{F}_{pf} + \nabla \cdot \left(\varepsilon_f \boldsymbol{\tau}\right) + \rho_f \varepsilon_f \mathbf{g} \tag{4}$$

where $\mathbf{u}$, ρ_f and p are the fluid velocity, density and pressure, respectively. $\mathbf{F}_{pf}$ is the volumetric particle-fluid interaction force in a computational CFD cell of volume, ΔV. $\boldsymbol{\tau}$ and ε_f are the fluid viscous stress tensor and local porosity.

Details regarding the governing equations and the equations to calculate the interaction forces and torques can be referred to [20, 21]. The bed geometry used in this work is a slot model with periodic boundary conditions applied to the front and rear direction. For such a geometry, 2D CFD and 3D DEM are employed as used elsewhere [20]. The bed is produced by generating particles randomly. Particles are generated randomly at different heights and allowed to come to rest to form a bed [21]. The bed is then fluidized by gas from central orifice at the bottom which is of size 4 CFD cells (0.001 m). Table 1 lists the parameters including the bed geometry, physical properties of particles and fluid used in the simulation.

Table 1. Various parameters used in simulation

Variable	Values
Bed geometry	
Bed width (x)	0.02 m
Bed height (z)	0.08 m
Bed thickness (y)	0.0004 m
Orifice width	4 CFD cells

CFD cells (x, z)	320 x 80 cells
CFD cell size (Δx , Δz)	0.00025 m
Particle Properties	
Shape	spherical
U_{mf}, m/s	0.0082
Number of Particles, N_p	200,000
Particle size, d_p	100 μm
Particle density, ρ_p	2450 kg/m^3
Particle-particle sliding friction, μ_s	0.4
Particle-wall sliding friction, μ_r	0.4
Normal and tangential damping co-efficient, c_n	0.3
Young's modulus, E	1 x 10^7 Pa
Poisson ratio, ν	0.3
Time step, Δt	1.6 x 10^{-6}
Gas properties	
Density	1.205 kg/m^3
Viscosity	1.8 x 10^{-5} Pa s

3. Results and discussion

Figure 1 shows the snapshots of the bubble formation process for a central jet velocity of 0.25 m/s. It can be observed that after the introduction of a central air jet to the bed, series of bubbles are formed one after another, the first bubble leaves the distributor at 0.043 s. The bubbles are formed at a certain distance above the distributor and rise through the bed before erupting at the top of the bed. This is consistent with general understanding of this process and the literature reported. For example, according to Tsukada and Horio [22], the jet behavior is transient and bubble formation is periodic in nature and it starts with the gradual formation of the bubble-like void at the top of the jet.

The previous studies have shown that after the formation of the first bubble, the air through the central jet is not able to form the next bubble immediately. Instead, an elongated void or jet is formed at the distributor [15] which merges with the bubble above, similar jet behavior is observed as shown in Figure 1 (t = 0.071 s). In the present work, since series of bubbles are formed, several such instances are recorded (Figure 1, t = 0.0126 s, t = 0.212 s, 0.343 s, etc). Therefore, the bubbles are categorised as two types: primary/ leading bubbles and trailing bubbles as shown in Figure 2. The primary bubbles are defined as the bubbles whose motion are not affected by the bubbles ahead them but they affect the succeeding bubbles. While the trailing bubbles are those bubbles whose motion is affected by the primary bubbles. For example, Bubble 1 (B1) is a primary bubble and B2 is a trailing bubble, similarly, B3 is a primary bubble and B4 and B5 are trailing bubbles (Figure 1 and 2) etc. The trailing bubbles catch up with the leading bubbles at some height in the bed thereby coalescing. The coalescence observed for the bubbles formed mainly consists of a leading-trailing type of coalescence [23]. Leading-trailing coalescence describes the scenario where two bubbles form consecutively rather than simultaneously. As they rise, the first bubble slows due to its interaction with the trailing bubble. The trailing bubble will then 'catch' and coalesce with the leading bubble. The coalesced bubble will move up the remaining height of the bed as a single bubble or may coalesce with other bubbles. Figure 2 plots bubble nose position against time. Several process of bubble formation together with bubble coalescence is depicted. There were three positions where coalescence occurred as shown in Figure 2. The first position is near the distributor just after bubble detachment at a bed height of 0.005-0.008 where after the detachment of bubble, and the bubble is sequentially followed by a jet which merges to form a larger bubble. The second coalescence occurs at a bed height of 0.01-0.015 m where a leading bubble was caught up by a trailing bubble as shown. The final coalescence occurs near the top of the bed at bed height of > 0.022 m where the merged bubbles erupt concurrently. The bubble breakup is not so significant since at lower height, the bubble recoalesces after bubble splits, and at the top of the bed, the bubble breaks up to erupt.

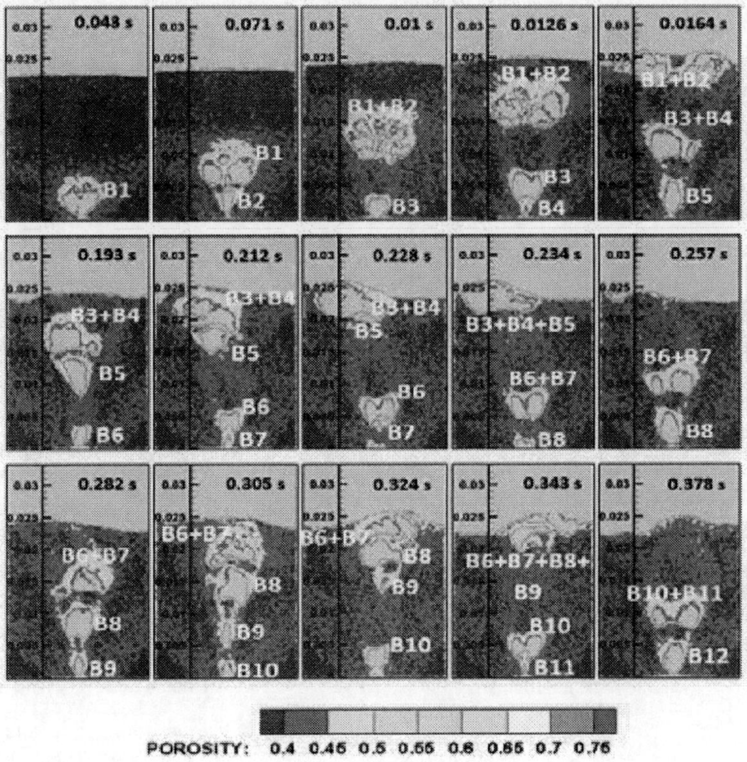

Figure 1: Bubble formation and trajectory snapshots for jet velocity of 0.25 m/s.

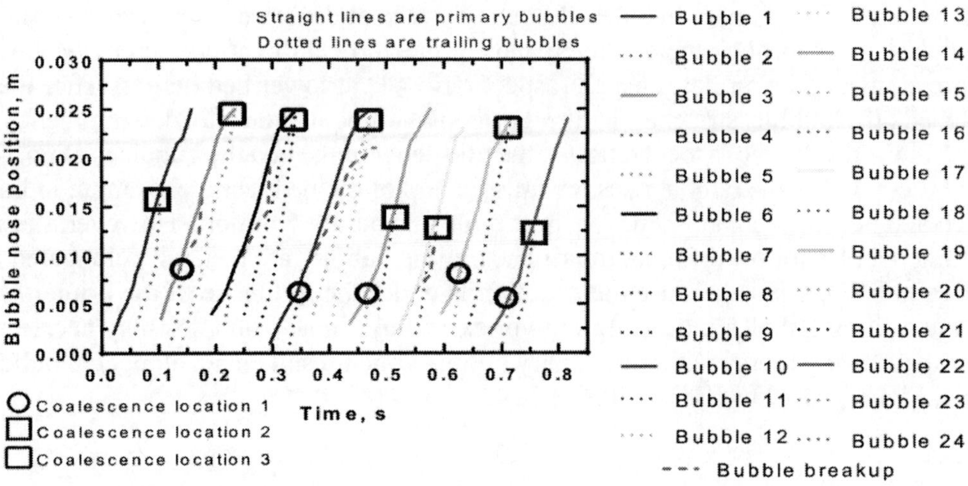

Figure 2: Motion of chain of bubbles for jet velocity of 0.25 m/s.

Figure 3 shows the bubble diameter at different positions. Figure 3a, b and c shows the change in bubble diameter with bubble location for both primary and trailing bubbles. It can be seen that the increase in the bubble diameter is because of the bubble coalescence while the decrease in bubble diameter is due to bubble splitting. In this process the size of the primary bubble keeps on increasing while the size of the trailing bubble decreases since trailing bubble is coalescing with the primary bubble. Figure 3d shows the change in bubble diameters of 3 primary bubbles at different bubble position. It can be seen that the bubble diameter and its change for the primary bubbles at different location are almost alike indicating the growth of primary bubbles are similar.

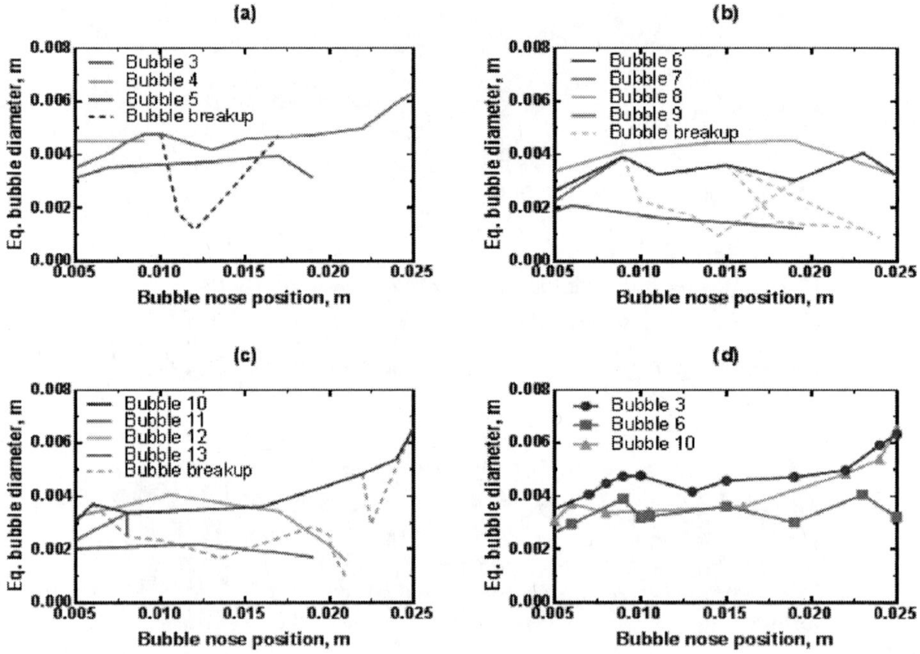

Figure 3: Bubble diameter at different positions.

Figure 4 shows the bubble shape characterized by bubble aspect ratio for different position of the bubble. The figure clearly shows the differences in the bubble shape for the primary and trailing bubbles. The primary bubbles although formed at different times are of similar shape. As the bubble rises through the bed it expands in the vertical direction thereby becomes more and more elongated i.e. prolate shaped (aspect ratio >1) at lower bed height. After the bubble is detached, the bubble stretches in the horizontal direction due to development of wake resulting in the decrease of aspect ratio. At the mid-level of the bed the bubble is almost circular (aspect ratio = 1). As the bubble rises towards the top of the bed it stretches more in horizontal direction and becomes oblate shaped (aspect ratio < 1) before eruption. However, the shape of the trailing bubbles follow different trend. The trailing bubbles are generally elongated in shape (aspect ratio >>1), this is because their motion is under the influence of the primary bubbles. When the trailing bubbles move into the vortex of the primary bubbles they accelerates and catches up to finally merge with the primary bubble which results in an elongated bubbles (For example, Figure 1 0.193 s (B5), 0.305 s (B8)).

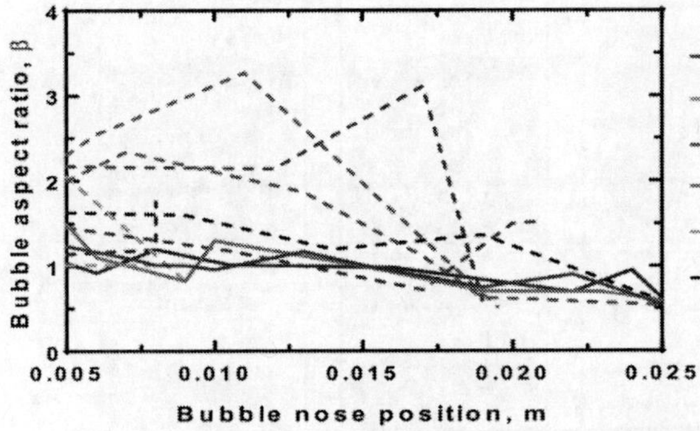

Figure 4: Bubble shape at different positions.

Figure 5 shows the bubble velocity for different position of the bubble. It is evident from the figure that the bubble velocity is distinctly different for primary and trailing bubbles. The bubble velocity for the primary bubble initially increases as it rises through the bed then remains constant (at mid-level of the bed) until it falls rapidly at the top of the bed during eruption. On the contrary, the trailing bubbles have very high bubble velocity around 1.5-2 times the velocity of primary bubbles. This is because the trailing bubbles are in acceleration to catch up and merge with the primary bubble which is at higher level.

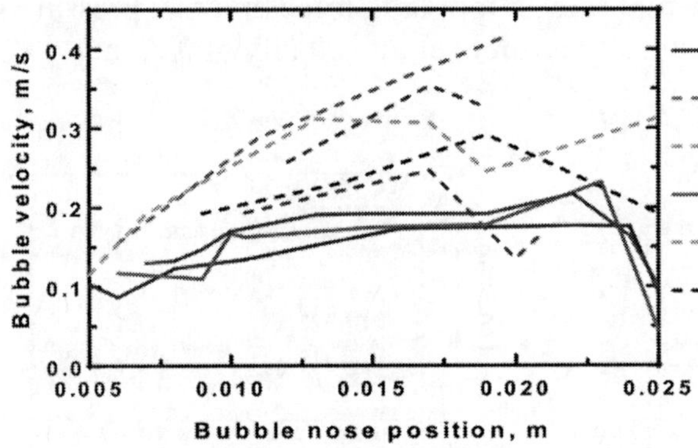

Figure 5: Bubble velocity at different positions.

One of the important forces in the process of fluidization is the interaction between particles and interstitial fluid which consists of the fluid drag force and pressure gradient force. This forces vary spatio-temporally. As the example, the spatial distribution of various forces (top - drag force; middle - pressure force; and bottom - particle-fluid interaction force) at different times is shown Figure 6. It should be noted that the force ratio is used by normalizing forces with particle gravity *mg*. It can be observed that these forces around gas bubbles are significantly different from the other areas. The drag force, pressure gradient force and their sum the particle fluid interaction force are larger in the dense areas around the gas bubbles than in the loose areas in the bubbles or far from the bubbles due to the differences in porosity. These forces are maximum in between the bubbles and minimum in the bubbles.

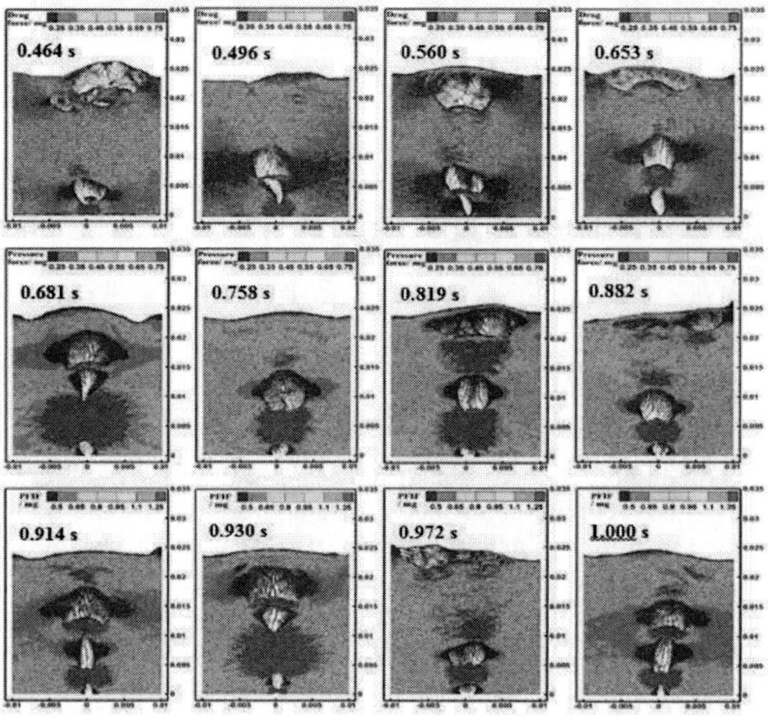

Figure 6: Spatial distribution of different forces (top: drag force, middle: pressure force and bottom: particle fluid interaction force) at jet velocity of 0.25 m/s.

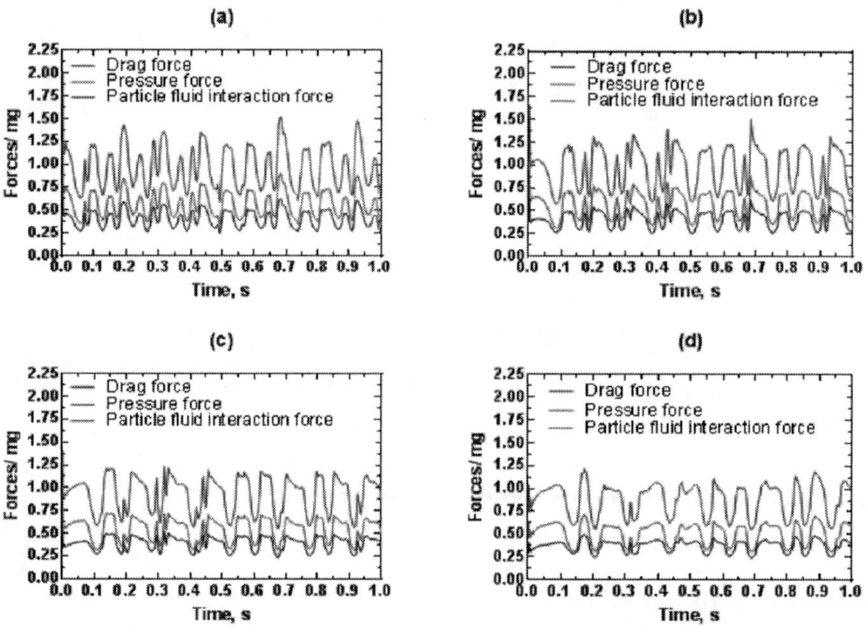

Figure 7: Different forces acting on the bed at different times for different bed height (axial direction): a) H = 0.005 m, b) H =0.01 m, c) H = 0.015 m and d) H = 0.02 m.

Figure 7 shows the temporal variation of drag force, pressure gradient force and particle fluid interaction force at different bed heights. It can be seen that the pressure gradient force is greater than the drag force at any level of the bed. Moreover, the magnitude of the forces decreases as the bed height increases which is due to the loss in the energy retained by the bubbles as the bed height increases. In addition, the peak depicted in the figures are due to the bubble passage. The maximum represents when the bubble has passed the point while the minimum represents when the bubble centroid is at that point. It can be observed that the peak gets plateaued as the bed height increases. This is because of the bubble coalescence. At lower level of bed there is no significant bubble coalescence, however as the bed level increases there is significant bubble coalescence present. Therefore, the gap in between the bubble increases leading to increased presence of solid material with less porosity thereby increasing the magnitude of the relative forces resulting in the formation of plateau as the bed height increases.

Conclusion

Formation of a series of bubbles in a fluidized bed operated with a single continuous jet was investigated using CFD-DEM to study the bubble characteristics such as bubble size, shape and velocity. The bubbles are categorised as two types: primary/leading bubbles and trailing bubbles. The primary bubbles are defined as the bubbles whose motion are not affected by the bubbles ahead them but they affect the succeeding bubbles. The trailing bubbles are those bubbles whose motion is affected by the primary bubbles. The trailing bubbles catch up with the leading bubbles at some height in the bed thereby coalescing. The size of the primary bubble increases at the cost of trailing bubbles. Both shape and velocity of the trailing bubbles are significantly different compared to primary bubbles. Trailing bubbles are elongated in shape and have higher bubble velocity. The force analysis showed that the drag force, pressure gradient force and their sum the particle fluid interaction force are higher in the dense areas in between the gas bubbles than in the bubbles or far from the bubbles.

References

[1] M. Rüdisüli, Characterization of rising gas bubbles in fluidized beds by means of statistical tools, PhD. Thesis ETH Zurich (2012).

[2] A. Busciglio, G. Vella, G. Micale, L. Rizzuti, Analysis of the bubbling behaviour of 2D gas solid fluidized beds: Part I. Digital image analysis technique, Chem. Eng. J., 140 (2008) 398-413.

[3] A.D. Lotfabad, S. Movahedirad, M. Sadeghi, Bubble Formation on a Single Orifice in a Gas Solid Fluidized Bed Using Digital Image Analysis, Iran. J. Chem. Eng., 13 (2016) 61.

[4] D. Harrison, L. Leung, Bubble formation at an orifice in a fluidized bed, Chem. Eng. Res. Des., 39 (1961) 409-414.

[5] H.S. Caram, K.-K. Hsu, Bubble formation and gas leakage in fluidized beds, Chem. Eng. Sci., 41 (1986) 1445-1453.

[6] K. Nosrati, S. Movahedirad, M.A. Sobati, A.A. Sarbanha, Experimental study on the pressure wave attenuation across gas-solid fluidized bed by single bubble injection, Powder Technol., 305 (2017) 620-624.

[7] Z. Amiri, S. Movahedirad, M. Shirvani, Particles mixing induced by bubbles in a gas - solid fluidized bed, AIChE Journal, 62 (2016) 1430-1438.

[8] J. Kuipers, W. Prins, W. Van Swaaij, Theoretical and experimental bubble formation at a single orifice in a two-dimensional gas-fluidized bed, Chem. Eng. Sci., 46 (1991) 2881-2894.

[9] J. Nieuwland, M. Veenendaal, J. Kuipers, W. Van Swaaij, Bubble formation at a single orifice in gas-fluidised beds, Chem. Eng. Sci., 51 (1996) 4087-4102.

[10] R.P. Utikar, V.V. Ranade, Single jet fluidized beds: experiments and CFD simulations with glass and polypropylene particles, Chem. Eng. Sci., 62 (2007) 167-183.

[11] M. Lungu, J. Wang, Y. Yang, Numerical simulations of flow structure and heat transfer in a central jet bubbling fluidized bed, Powder Technol., 269 (2015) 139-152.

[12] Y. Zhang, R. Xiao, M. Ye, Z. Liu, A numerical study of the bubble induced pressure fluctuation in gas-fluidized beds, Powder Technol., 314 (2017) 387-399.
[13] D. Gera, M. Gautam, Y. Tsuji, T. Kawaguchi, T. Tanaka, Computer simulation of bubbles in large-particle fluidized beds, Powder Technol., 98 (1998) 38-47.
[14] O.O. Olaofe, M.A. van der Hoef, J.A.M. Kuipers, Bubble formation at a single orifice in a 2D gas-fluidized bed, Chem. Eng. Sci., 66 (2011) 2764-2773.
[15] J.K. Pandit, X. Wang, M. Rhodes, A DEM study of bubble formation in Group B fluidized beds with and without cohesive inter-particle forces, Chem. Eng. Sci., 62 (2007) 159-166.
[16] L.-w. Rong, J.-m. Zhan, C.-l. Wu, Effect of various parameters on bubble formation due to a single jet pulse in two-dimensional coarse-particle fluidized beds, Adv. Powder Technol., 23 (2012) 398-405.
[17] S. Shrestha, Z. Zhou, A CFD-DEM study of single bubble formation in gas fluidization of spherical and non-spherical particles, EPJ Web of Conferences, EDP Sciences, 2017, pp. 15026.
[18] Z. Amiri, S. Movahedirad, M. Shirvani, O. Vahidi, The role of bubble injection characteristics at incipient fluidization condition on the mixing of particles in a gas-solid fluidized bed at high operating pressures: A CFD-DPM approach, Powder Technol., 305 (2017) 739-747.
[19] L. Tan, I. Roghair, M. van Sint Annaland, Discrete particle simulations of bubble-to-emulsion phase mass transfer in single-bubble fluidized beds, Particuology, 33 (2017) 80-90.
[20] Z. Zhou, D. Pinson, R. Zou, A. Yu, Discrete particle simulation of gas fluidization of ellipsoidal particles, Chem. Eng. Sci., 66 (2011) 6128-6145.
[21] J. Gan, Z. Zhou, A. Yu, CFD–DEM modeling of gas fluidization of fine ellipsoidal particles, AIChE Journal, 62 (2016) 62-77.
[22] M. Tsukada, M. Horio, Gas motion and bubble formation at the distributor of a fluidized bed, Powder Technol., 63 (1990) 69-74.
[23] K. Lim, J. Zhu, J. Grace, Hydrodynamics of gas-solid fluidization, International journal of multiphase flow, 21 (1995) 141-193.

504801 Statistical Analysis on Large-Scale Direct Numerical Simulation of Gas-Solid Flow

Abstract:

Particle-resolved direct numerical simulation (DNS), where the computational grid for the fluid phase is reduced to below the particle size, the flow around each particle is fully resolved, and the fluid-solid interaction force is directly obtained by integrating the viscous force and pressure on the particle's surface. It has been regarded as the most accurate numerical method for simulation of gas-solid flow. However, the main disadvantage is its huge computational cost resulting from small grid size and time step limited by Kolmogorov length and time scales. Only hundreds of particles scale is reported for DNS of gas-solid flow in the latest literature, which is very different from the number of particles in the real gas-solid flows. Therefore, small-scale simulation results are hardly to be established a true and reliable constitutive law for gas-solid flow. In order to solve the problem of computational speed and scale, we use the lattice Boltzmann method (LBM) with good parallelism to replace the traditional N-S equation to describe the gas phase, an immersed boundary method in framework of LBM has been adopt to realized the fluid-solid coupling to avoid a stair-step representation of the solid particles' surfaces (Wang et al., 2010; Zhou et al., 2011), and the multi graphics processor units (GPUs) parallel computing is also implemented. Taking advantage of the inherent parallelism of LBM and the attractive Flops/Price ratio of GPU, we have implemented 576 GPUs parallel computing on a Mole-8.5 system and conducted the largest scale DNS of gas-solid suspensions so far, with 1,166,400 solid particles in an area of 11.5cm x 46cm for a two-dimensional system and 129,024 solid particles in a domain of 0.384cm x 1.512cm x 0.384cm for a three-dimensional system (Xiong et al., 2012). The scale of DNS data has been reached the size in traditional computational grid, which implies the really meaningful statistical results from large-scale DNS of gas-solid flows were obtained for the first time. The effects of mesoscale structure on the interaction force between gas and solid phases (Zhou et al., 2014) and the statistical properties of particles (Liu et al., 2017) were explored, which may provide the corresponding constitutive relation and detailed microscopic information for discrete particle simulation and two-fluid model.

References

1. Limin Wang, Guofeng Zhou, Xiaowei Wang, Qingang Xiong, Wei Ge. 2010. Direct numerical simulation of particle-fluid systems by combining time-driven hard-sphere model and lattice Boltzmann method. Particuology, 8: 379-382.
2. Guofeng Zhou, Limin Wang, Xiaowei Wang, Wei Ge. 2011. Galilean-invariant algorithm coupling immersed moving boundary conditions and Lees-Edwards boundary conditions. Physical Review E 84, 066701.
3. Qingang Xiong, Bo Li, Guofeng Zhou, Xiaojian Fang, Ji Xu, Junwu Wang, Xianfeng He, Xiaowei Wang, Limin Wang, Wei Ge, Jinghai Li. 2012. Large-scale DNS of gas–solid flows on Mole-8.5. Chemical Engineering Science 71: 422–430.
4. Guofeng Zhou, Qingang Xiong, Limin Wang, Xiaowei Wang, Xinxin Ren, Wei Ge. 2014. Structure-dependent drag in gas-solid flows studied with direct numerical simulation. Chemical Engineering Science, 116: 9-22.
5. Xiaowen Liu, Limin Wang, Wei Ge. 2017. Meso-scale statistical properties of gas–solid flow—a direct numerical simulation (DNS) study. AIChE Journal 63:3-14.

Copyright © American Institute of Chemical Engineers. All rights reserved.

509635 Capillary Interaction in DEM Simulations of Wet Particulate Materials

Abstract:

Processing stages such as granulation, encapsulation, coating, drying, particle melting are present in a variety of industries, including the manufacture of food, pharmaceuticals, catalyst, mechanical part products. The presence of some liquid between the particles determines forces and complex particle formation or breakage whose understanding and characterization is far from complete. Discrete Element Method (DEM) simulations possess the ability to carry out analysis of dense particulate material under actual industrial processing conditions, being able to include interaction contributions at the individual particle level.

In the present work, microscopic models of particle-particle capillary interaction for inclusion in advanced DEM simulations are investigated. Liquid bridge formation, interaction force, stability and rupture are investigated as a function of the liquid volume, liquid wetting properties (e.g. contact angle), particle separation distance, particle size ratio. The extensive set of results are obtained adopting a rigorous computation of the equilibrium bridge shape and characteristics. Comparison with existing approaches, such as the toroidal approximation, will be finally presented.

Copyright © American Institute of Chemical Engineers. All rights reserved.

508789 An Idea of DEM-CFD Coupling Model Decoupling Spatial Averaging Scale from Computational Cell Size

Abstract:

DEM-CFD coupling model is widely used for the numerical simulation of dense gas-solid flows. Because it is a mesoscopic approach, fluid motion is obtained by solving spatially-averaged continuity and momentum equations. It is usual that computational cell size used for CFD calculation is same with the spatial-averaging scale of governing equations and the results obtained by the model can change depending on the computational cell size. Fine computational cell is desired for accurate CFD calculation, on the contrary, sufficiently large averaging scale is required to fulfil the requirement of spatial averaging. These are contradictory and it is important to adopt a proper cell size carefully to obtain reasonable results. It could be a big problem in the case particles with different size exist and particle size changes over time and position.

In the present study, an idea of DEM-CFD coupling model that decouples spatial averaging scale from calculation cell size used for CFD calculation is discussed. We present the concept of proposed model first and develop discussions on its feasibility and validity.

Copyright © American Institute of Chemical Engineers. All rights reserved.

508739 A Fundamental Framework for Simulating Large-Scale Particulate Systems Using the Direct Simulation Monte Carlo Method

Abstract:

Abstract: A hybrid direct simulation Monte Carlo (DSMC) and multiphase particle in cell (MPPIC) scheme is developed to accurately simulate large-scale particulate systems. The MPPIC method tracks parcels that represent a large number of real particles, and can be used to simulate large-scale particulate processes with complex particle-scale physics. Despite the efficiency of the MPPIC method, its usefulness for dynamic systems is limited due to evaluation of the solids stresses and solids heat flux. The DSMC is a fundamental modeling approach that is used to calculate collisions between parcels. In doing so, the DSMC method effectively evaluates the collision integral in the Enskog equation. The DSMC method therefore does not require simplifying assumptions that are often used to derive continuum closures, such as closeness to equilibrium and small Knudsen numbers. The validity of the Enskog equation breaks down as the solids phase reaches the packed limit. Consequently, pure DSMC simulations show errors such as over-packing in dense regions. To eliminate errors associated with over-packing, the MPPIC solids stress model is used when the solids concentration is close to the packed limit and the DSMC method is used for dilute and moderately dense regions. Simulations using the hybridized MPPIC-DSMC method are completed for a homogeneous cooling system, fluidized bed, and rider flow. The new hybrid simulation technique is compared to more traditional methods, such as two fluid modeling, MPPIC, and the discrete element method.

Copyright © American Institute of Chemical Engineers. All rights reserved.

506236 Comparison of CFD Models in Predicting the Fluidization Behavior of Geldart B Particles

Abstract:

A particular flow regime occasionally observed when fluidizing Geldart B particles in small scale units is slugging. This flow regime, characterized by high pressure fluctuations and big bubbles that can cover the entire section of the fluidizing column (square-nosed slugs), may result in difficulty of extrapolation of the hydrodynamic results obtained for small units into larger scales.

CFD models represent a potential tool for the design of gas-fluidized beds. Despite the efforts of the scientific community, the ability of these models in predicting the hydrodynamic behavior of these multiphase systems is still limited. Robust and reliable CFD tools that can predict the hydrodynamics of group B particles slugging in lab or pilot scale units is extremely useful for beforehand definition of the operating conditions and geometry that can assure cold flow model experimentation reproducing the same hydrodynamics of large scale reactors.

In this work, two different commercial CFD tools have been compared in their ability of predicting hydrodynamic features of slugging gas-solid fluidized beds. The study is of special interest to chemical looping combustion process that employs Geldart B particles but also to fluidized bed combustion or gasification processes. Different parameters such as bed expansion, pressure fluctuations, voidage and solid velocity obtained experimentally have been compared with CFD results and the effects of simulation parameters on the predicted flow structures have been discussed.

Copyright © American Institute of Chemical Engineers. All rights reserved.

507540 Evaluation of Drag Models for Gas-Solid Fluidization of Geldart a Particles in All Flow Regimes

Abstract:

Coarse grid two-fluid simulation of gas-solid fluidized bed reactors based on the kinetic theory of granular flow exhibits a significant dependence on drag models, especially for Geldart A particles. Inappropriate selection of drag models leads to inaccurate fluidization behavior predictions. This study focused on the evaluation of drag models derived from different methods for 3D two-fluid model simulations of gas-solid fluidization of Geldart A particles in bubbling fluidization, turbulent fluidization, fast fluidization and dilute phase transport regimes. Eight candidate drag models were selected, which included five heterogeneous drag models and three homogeneous drag models that belong to six types. Comparison with available experimental data clearly demonstrates the need for modification of homogeneous drag models to account for the effect of mesoscale structures (bubble and cluster). The Sarkar drag model and EMMS drag models are found to achieve superior predictions in all fluidization regimes, while the other drag models are found to be capable of predicting fewer fluidization regimes. The results of this work provide a guideline for choosing appropriate drag models for simulating Geldart A particles and provide suggestions on developing more reliable and general drag models applicable in all flow regimes.

Copyright © American Institute of Chemical Engineers. All rights reserved.

514089 The Vagaries of Granular and Particle Laden Flow

Abstract:

In the 80s, the development of the kinetic-collisional theory for granular flow (KTGF) represented a major breakthrough in the modelling of the hydrodynamics of particulate flow: based on the analogy between solid particles and gas molecules, a kinetic theory was proposed which produced constitutive equations for rapid granular flow when binary, instantaneous and inelastic particle collisions occur. In practical situations, however, the flow of particles presents complexities that escape successful description based on the KTGF alone. Many complicated phenomena happen at the meso-scale, the scale between "small" and "large": as an example, clusters and agglomerates are formed at a length scale between particle size and equipment size. In addition, particle-particle interactions, other than inelastic and instantaneous collisions, might occur, and more complex theories are needed to capture the observed behaviour.

The talk will describe the journey from the first rigorous effort to model granular and particle laden flow to recent theories and models to explain the vagaries observed in practical situations. Open problems, in need of a solution, will be presented and the search for a "unifying" theory will be discussed.

Copyright © American Institute of Chemical Engineers. All rights reserved.

514090 Scaling Up Particulate Process - Lessons from Industry

Abstract:

In the 1980â€™s, the Rand Corporation identified challenges in the commercialization of processes involving particles. Particulate process plants took longer to start up and achieved lower production rates than expected. Not enough has changed in more than 30 years. Processes involving particles are inherently more difficult than those for liquids and gases. A consistent lack of fundamental understanding dooms industry to empirical approaches, and the sensitivity of experiments to scale and equipment configuration leads to high costs and long development periods. Modeling by the discrete element method (DEM) has not yet fulfilled its promise, especially for small particles. This presentation will provide examples of the challenges involved in some of the unit operations of particle technology, and discuss strategies for scale-up, including the use of equipment suppliers, toll processors, modeling and pilot plants to bridge the knowledge gaps. Even with such strategies, education of both the engineer and the business leader is necessary to manage expectations and appropriately allocate resources.

Copyright © American Institute of Chemical Engineers. All rights reserved.

510390 The Divide between Academia and Industry in Modeling Gas-Solids Reacting Flows

Abstract:

Developing new and efficient commercial scale multiphase reactors is extremely difficult as one needs to consider the complex interactions over a wide range of both temporal and spatial scales encountered in these systems (from molecules to macroscale). Therefore, it is important to codify and automate the knowledge collectively acquired so far, reserving human resources for the creative solutions that build upon codifying past learnings. Computational science (algorithms, theory and modeling, computer science, etc.) combined with exponential growth in computing hardware has great potential to provide us valuable tools for improving the effectiveness of existing plants but also design new reactors. Industry is in desperate need of fast simulation tools that are comparable to the accuracy of the single-phase reacting flows to advance the adoption of multiphase flow reactors and primarily to reduce the risk of scale-up. On the other hand, academia and national laboratories are focused on improving numerical methods, submodels or certain physics to push the state-of-the-art and predictability of the models. However, there is tremendous need for rapid integration of various capabilities that exploit latest computer hardware improvements, robust model reduction techniques, etc. that would improve the computational tools available to the industry to deliver the next generation of efficient and cost-effective reactors. These developments typically fall into the valley of death where industry and software vendors feel these efforts are far-fetched while the academic world considers it incremental. One way to address this divide is to develop mechanistic upscaling framework to coarse-grain information from more detailed modeling approaches to fast models accessible to the industry with relatively minimal information loss.

I will present such an approach using a gas-phase fluidized bed polymerization reactor as an example. First, I will provide an overview of the various models currently used at the different scales in gas-solids reacting flows. The coupling across the scales will be introduced through few different approaches:

1. Discrete-Continuum Coupling using Discrete Element Method for particles
2. Upscaling data from Discrete Element Modeling results to continuum based Computational Fluid Dynamics (CFD)
3. Upscaling information from CFD to simplified reactor network models

In conclusion, the presentation will elucidate the approaches and opportunities in bridging the gap between academic advances and industry needs to model and advance our understanding of gas-solids reacting flow reactors.

Copyright © American Institute of Chemical Engineers. All rights reserved.

507029 A CFD-DEM Analysis of Particle Attrition in a Cyclone

Abstract:

Cyclones are used to separate entrained particles from the flue gas streams. Particles entering a cyclone are subjected to a radial centrifugal force field, driving them to the cyclone walls, where they experience collisional and rapid shearing stresses. For frangible particles this could lead to attrition and dust generation, whilst for hard particles wear of cyclone wall could occur.

In this work we address the issue of attrition of manganese di-oxide particles used in the CLC process. A CFD-DEM analysis of particle motion and interactions is used to calculate the frequency of collisions, impact velocities and shear stresses. These are then coupled with a model of particle breakage due to impact and of surface wear, in order to predict attrition in a cyclone. The outcome of the work is presented, providing a correlation of the attrition extent as a function of particle properties and operating conditions.

Copyright © American Institute of Chemical Engineers. All rights reserved.

DEM Simulation of Cylinders and Capsules in a Fluidized Bed

*Oleh Baran**
Thomas Eppinger
Kuanjin Han

Material presented at World Congress on Particle Technology in Orlando on April 23, 2018

Abstract

Fluidization of non-spherical particles of cylindrical or close to cylindrical shape is part of many industrial processes, especially in relation to biomass fuel production, fluidized bed gasifiers and pharmaceutical tablet coating processes. In these applications accurately accounting for the details of particle shape is important for safe and efficient design of industrial equipment. In this work the numerical model based on the Discrete Element Method (DEM) fully coupled with CFD is used to predict the effect of particle aspect ratio and presence of sharp edges in the particle's shape on the fluidized bed hydrodynamics. The model is implemented within commercial software Simcenter STAR-CCM+ version 12.06 developed by Siemens PLM. The model assumes particles are cylinders or spherocylinders (capsules). It accurately resolves the collisions between cylindrical particles or between capsules without employing conventional method of "spheres glued together" for non-spherical particles.

Cylindrical Particles

Capsules

The model is then used to simulate cylindrical particles or capsules of the same volume, mass, aspect ratio, and frictional properties in a cylindrical container with air inlet at the bottom and pressure outlet boundary at the top with the range of inlet velocities both below and above minimum fluidization velocity. The results for pressure drop, particle packing and particle preferred orientation are analyzed for both cylinders and capsules and compared to the correlations reported in the literature. The results for the percentage of particles aligned with fluid flow direction in bubbling flow regime indicate the presence of surprisingly longer timescale for establishing steady state preferred orientation of particles. The effect of the absence of the sharp edges in capsules compared to cylinders on the hydrodynamics of the fluidized bed is discussed in details.

Introduction

The computational models for the flow of elongated particles in fluidized bed can bring valuable insight in understanding the effect of particle shape on fluidization state inside industrial reactors and ultimately help designing cost effective fluidization processes and equipment. Notably Discrete Element Method (DEM) is recognized as promising computational approach for modeling the flow of non-spherical particles. For example, (Borzsonyi and Stannarius 2013), (Guo, et al. 2013) (Abreu, Tavares and Castier 2003) (Williams and Philipse 2003) use DEM to study interesting aspects on packing and shear resistance of rods without coupling to fluid flow and (Mema, et al. 2017), (Zhong, et al. 2009), (Mahajan, et al. 2017), (Cai, et al. 2016) use DEM coupled with CFD to investigate the flow of elongated particles in fluidized bed. Experimental data for validating DEM-CFD simulations of above systems are also becoming available recently, see, for example, (Buist, et al. 2017), (Liao 2013). However, DEM by itself and DEM-CFD coupling for non-spherical particles are known to be computationally too expensive to be applied to even smaller scale industrial scale problems.

In this work new DEM-CFD model is introduced for simulating cylinders or capsules in a fluidized bed which offers the computational speed advantages due to particular choice of modeling the contacts between non-spherical particles and applicability to complex geometries advantage due to special cell clustering approach in 2-way coupling between solid and fluid phases.

Contact model for arbitrarily shaped discrete elements

Resolving collisions between particles in the distinct feature of DEM, see Fig 1 below.

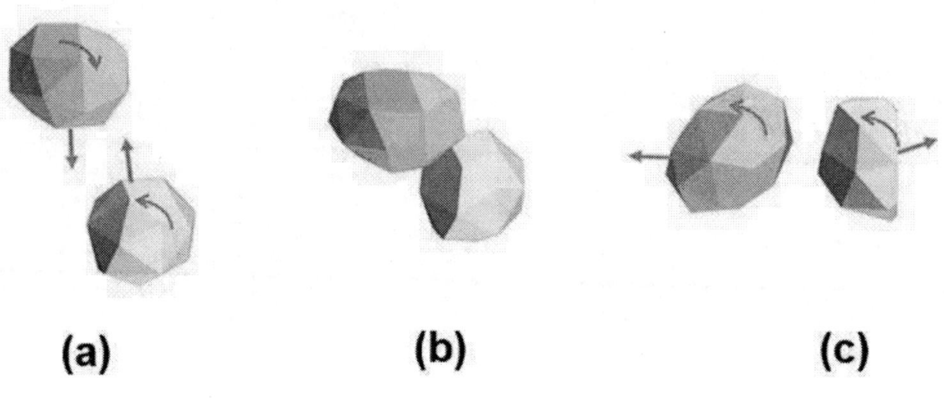

Figure 1. Collision between non-spherical particles. (a) Before collision. (b) During collision. (c) After collision. Blue arrows show particle's linear velocities and green arrows show particle's rotational velocities

Generalizing well validated and tested soft sphere approach for DEM on non-spherical particles is described next. Within this approach particle collision (see Fig 1 (b)) takes finite time or many DEM timesteps and can be described by the physics of spring-dashpot system as shown on schematics in Fig 2.

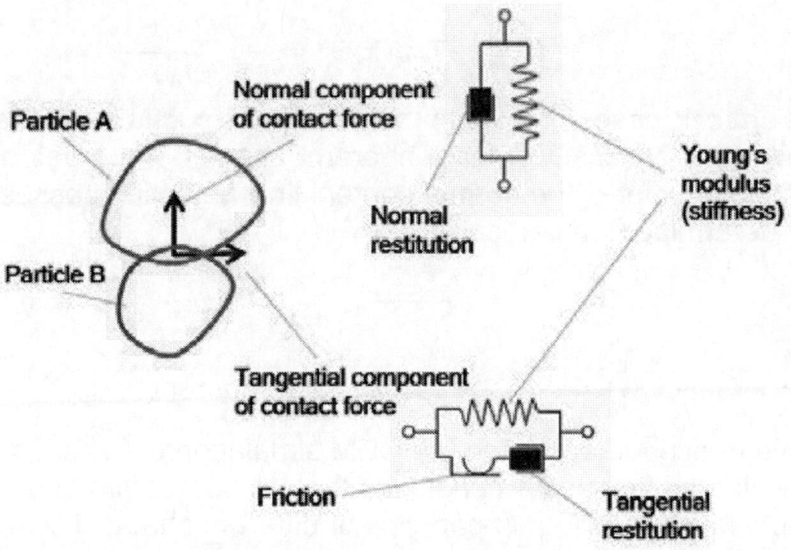

Figure 2. Schematics for soft collision approach in DEM

Within this new soft collision approach colliding Particle A and Particle B are rigid bodies that do not deform. During collision the overlap between these two arbitrary shape objects is monitored and contact force which depends on this overlap is calculated. Total contact force is a sum of normal and tangential components. According to (Feng, et al., 2012) the direction of the normal component can be uniquely determined and it provides the maximum rate of slowing down two impacting particles.

As shown in (Feng, et al., 2012) the elastic component of the normal contact force must conserve mechanical energy and, in order to do so, it must satisfy several requirements described next.

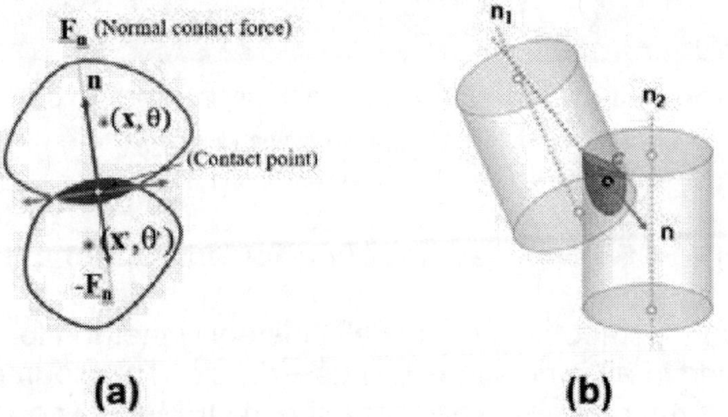

Figure 3. (a) Two colliding bodies with reference points. (b) Collision between axi-symmetric objects

In a fixed global coordinate system two points on the two colliding bodies have coordinates $\boldsymbol{x}$ and $\boldsymbol{x}'$ and the orientations of these two bodies are described in terms of variables Θ and Θ', see Fig. 3 (a). The normal elastic component of the contact force is the function of position and orientation of two bodies, $\boldsymbol{F}_n = \boldsymbol{F}_n(\boldsymbol{x}, \boldsymbol{x}', \Theta, \Theta', \dots)$ and it must obey the following "Energy Conservation Theorem". Contact model will be energy conserved if there exists a scalar potential function $\phi(\boldsymbol{x}, \Theta)$ such that

$$F_n = -\nabla_x \phi(x, \Theta) = -\frac{\partial \phi(x, \Theta)}{\partial x}$$

For more specific cases of axi-symmetric objects such as cylinders or capsules, see Fig. 3 (b), the following "Normal Contact Theorem" applies. For a pair of axi-symmetric bodies in contact with contact point c, the normal contact line n , which passes the two central lines, can be uniquely determined. In particular,

$$c = x + \frac{n \times M}{|F_n|} + \alpha n$$

Here $M = -\nabla_\Theta \phi(x, \Theta) = -\frac{\partial \phi(x,\Theta)}{\partial \Theta}$ is a contact moment.

With above described approach the DEM simulations of various shape particles were conducted and validated first under conditions that do not require coupling to fluid. One such validation example involves piling up particles of different shape. The final state of the simulated process is shown in Fig 4 below.

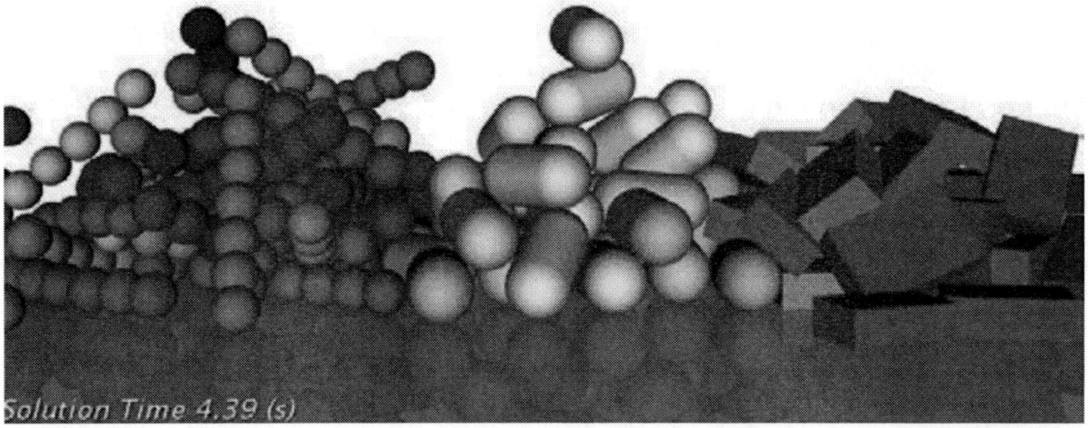

Figure 4. Interaction between different particle types. Left: blue clumped particles (flexible rods). Middle: Yellow capsules (spherocylinders). Right: Green polyhedral particles (boxes)

DEM-CFD two-way coupling for non-spherical particles

The novel source smoothing using cell clustering method for 2-way coupling between DEM and CFD is used in all simulations discussed next. This source smoothing algorithm considerably reduces the mesh size dependence in DEM-CFD coupling. This algorithm was first introduced in (Baran, Greening and Kodl, Numerical Simulations of Fluidized Beds in Fine-Grid Geometries Using Source Smoothing with Cell Clustering Method for DEM-CFD Coupling 2016) and (Baran, Eppinger, et al. 2016).

The drag force on capsules or cylinders is set to be a combination of Wen-Yu model and Ergun based model

$$\overrightarrow{F_{drag}} = \frac{1}{2}\rho_f C_d A_d |\vec{u} - \vec{v_p}|(\vec{u} - \vec{v_p}); \qquad C_d \text{ is a function of } Re_p = \frac{\rho_f |\vec{u} - \vec{v_p}| d}{\mu}$$

$$C_d = \begin{cases} \frac{4}{3}\left(\frac{150(1-\epsilon)^2}{\epsilon Re_p} + 1.75\right); & \text{if } \epsilon < \epsilon_{\min};\ Ergun \\ \frac{24 + 3.6 Re_p^{0.687}}{Re_p}\epsilon^{-3.65}; & \text{if } \epsilon \geq \epsilon_{min};\ Wen\,Yu \end{cases}$$

ϵ_{min} is the cutoff void fraction (=0.8)

A_d is a projected area of particle. For cylinders and capsules projected area is recalculated every DEM timestep as the orientation of particle relative to the fluid flow is continuously changing, see Fig. 5.

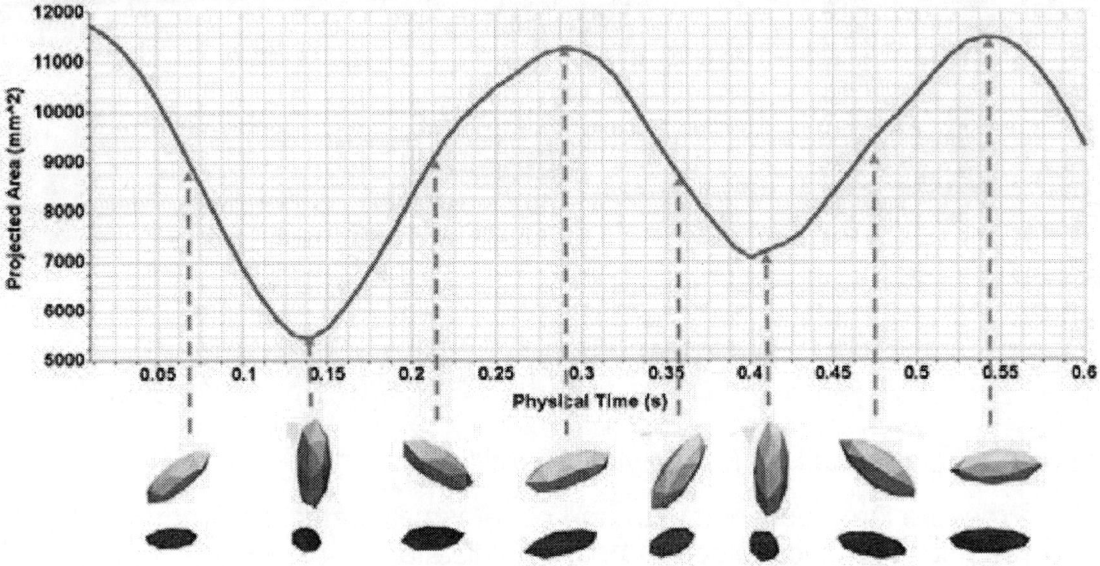

Figure 5. Projected area of elongated particle rotating in constant fluid flow coming from bottom to top. Orientation of particle for selected points in time are shown below the bottom axis along with the shape of projected area colored in black.

Fluidized bed setup

The simulations of capsules and cylinders of various aspect ratio in fluidized bed are discussed next. Geometry includes simple cylindrical container of diameter 0.8 m and height 2 m with velocity inlet at the bottom and pressure outlet at the top with gravity pointing downwards, see Fig. 6. Total simulated time is 20 seconds. The inlet velocity is zero for the first second during settling phase, then it is linearly ramped over the next four seconds to the final maximum inlet velocity $v_{inlet} = 4\,m/s$ and stays at this value until the end of simulation. In all simulated cases the number of particles is always 28320. The density of material particles are made from and their equivalent sphere diameter are fixed to have values $2200\ kg/m^3$ and $20\ mm$ respectively. This means regardless the shape of particle the total amount of solids is always the same. The shape and dimension of particles used in this study are given in Fig. 6. For the base case of capsules with aspect ratio 2 (short capsules) the maximum inlet velocity is 1.22 times minimum fluidization velocity as this study focuses mostly on the effect of particle shape on the hydrodynamics in a fluidized state regime.

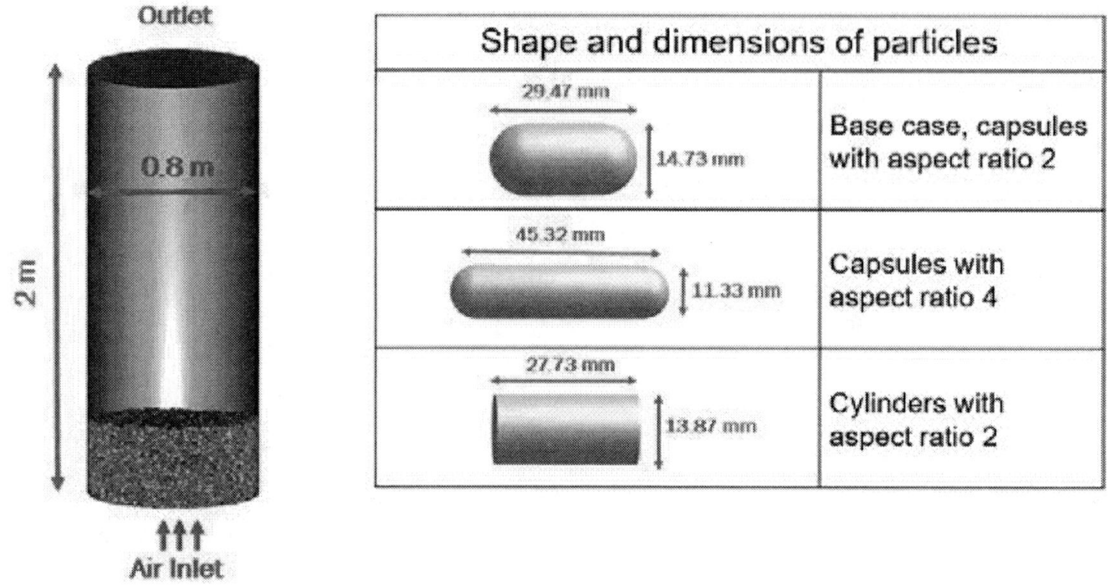

Figure 6. Fluidized bed setup and types of particles used in simulation.

All simulations are set to monitor the following quantities.

- Pressure Drop between inlet and outlet is calculated and compared with Ergun prediction for inlet velocities below fluidization
- Bed height is calculated as the vertical distance between inlet and center of mass of all particles times 2
- Void fraction is obtained by averaging in volume cells below fill level. It is used for evaluating the Ergun prediction for pressure drop for the inlet velocities below minimum fluidization velocity
- Alignment of cylinder's or capsule's orientation with the direction of fluid flow, see Fig. 7

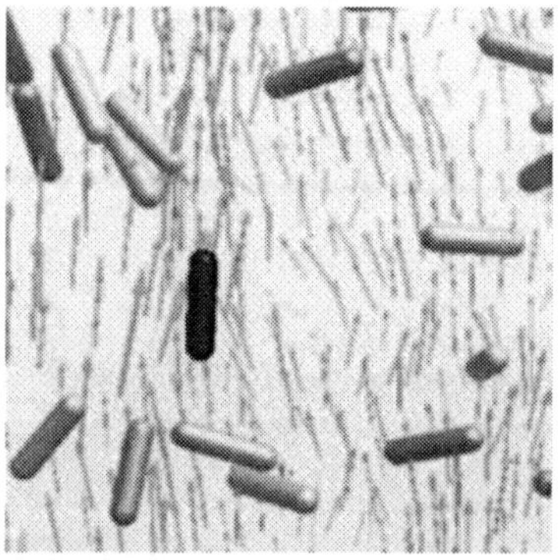

Figure 7. The fluid direction is shown by vectors. Particle aligned with fluid direction is shown in dark green

In particular, the angle between particle's main symmetry axis and local direction of fluid, angle of attack, is evaluated for every particle at the end of each fluid timestep and the status of "aligned" is set for each particle with angle of attack smaller than 25 degrees. Number of aligned particles or fraction of particles aligned with flow as a function of time is then available for analysis.

Simulation results

The snapshots for particle-fluid states in fluidized state for three shapes of particles are shown in Fig. 8

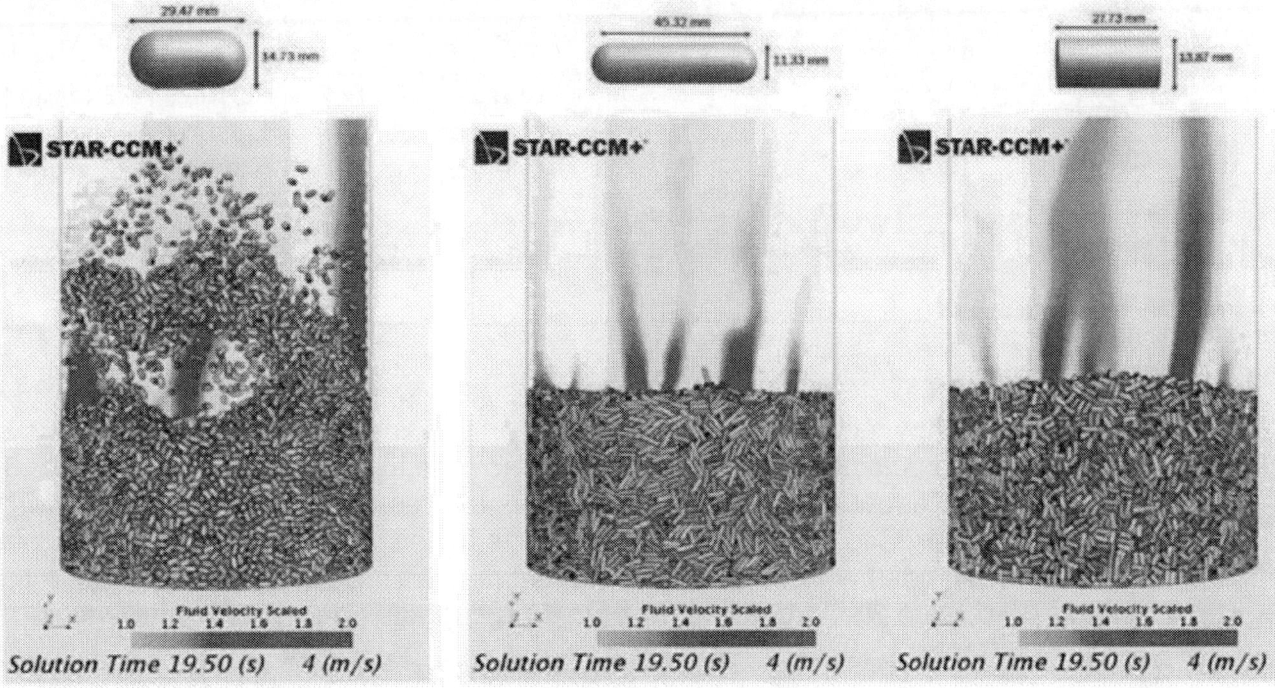

Figure 8. Rendering of particles in 3D (yellow) and fluid velocity magnitude (blue) on cross-section plane for cases of capsules with aspect ratio 2 (left), capsules of aspect ratio 4 (middle), and cylinders of aspect ratio 2 (right).

3D visualization of the magnitude of fluid velocity using resampled volume technology is shown in Fig. 9.

Figure 9. 3D rendering of fluid velocity for the case with capsules of aspect ratio 2

The pressure drop as a function of time for each considered case is shown in Fig. 10.

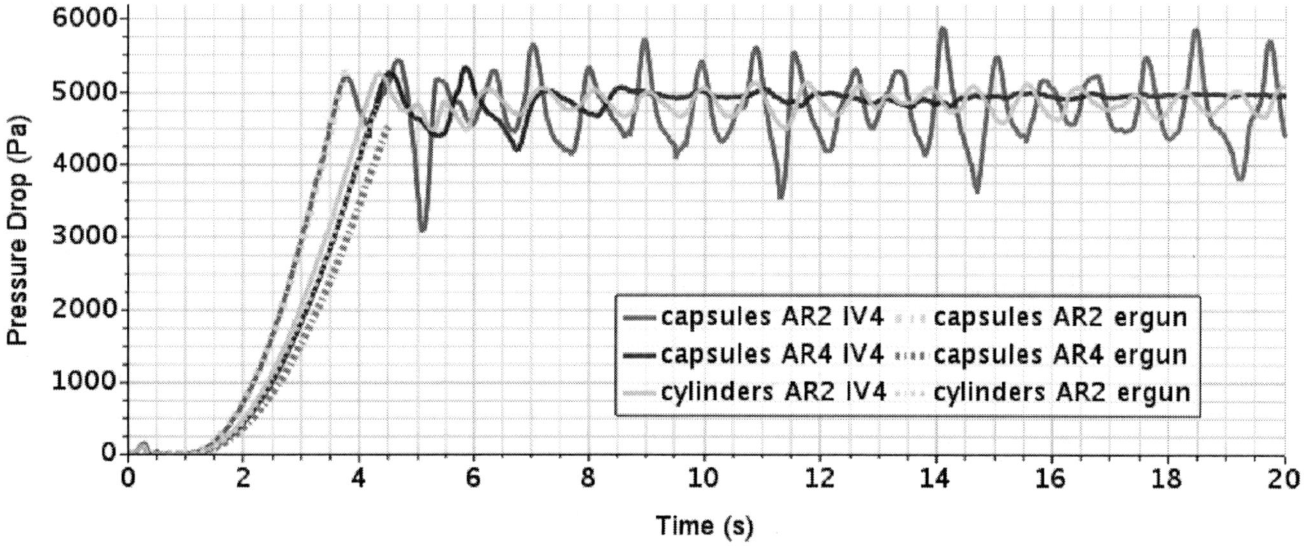

Figure 10. Pressure drop as a function of time

The results show good agreement with Ergun prediction for the inlet velocities below minimum fluidization velocities (which correspond to time interval between 1 s and 3.8 seconds). They also show much higher fluctuations of pressure for the case capsules with aspect ratio 2 compared to both other cases. This is consistent with visual analysis of the animations for each case where bubbling flow regime was prevailing for short capsules and channeling flow regime was prevailing for both long capsules and short cylinders.

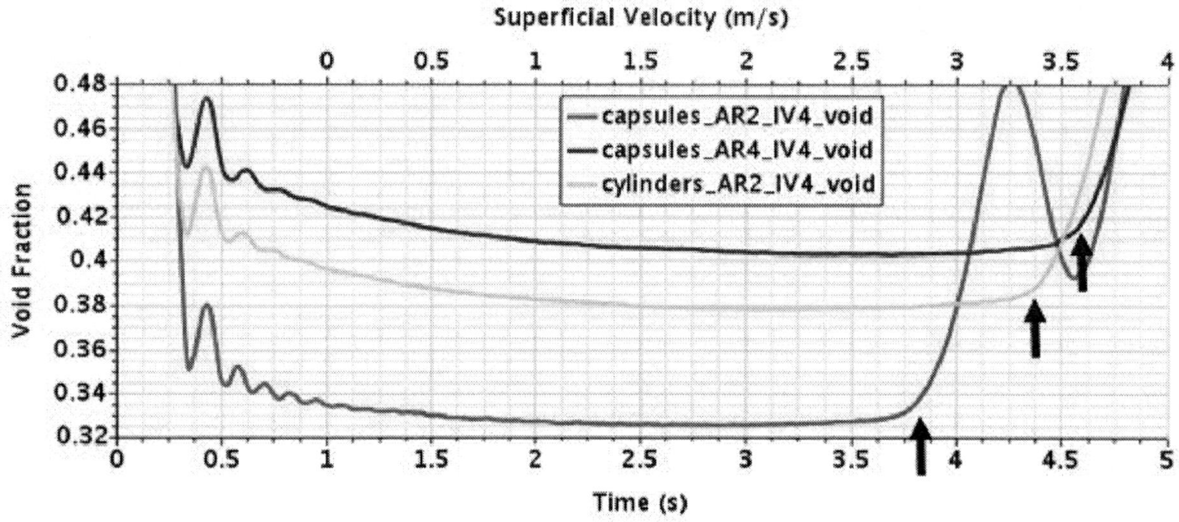

Figure 11. Void fraction below fill level as a function of time (bottom axis) and superficial velocity (top axis). Black arrows indicate the state of the onset of fluidization.

Fig. 11 shows the results for void fraction of packed particles. These results are used to estimate the minimum fluidization velocity as the threshold inlet velocity above which void fraction starts rapidly increasing. In particular, the minimum fluidization velocity is 2.7 m/s for short capsules, 3.5 m/s for long capsules and 3.2 m/s for short cylinders. The values of void fraction were also used in calculating the Ergun prediction for pressure drop plotted in Fig. 10.

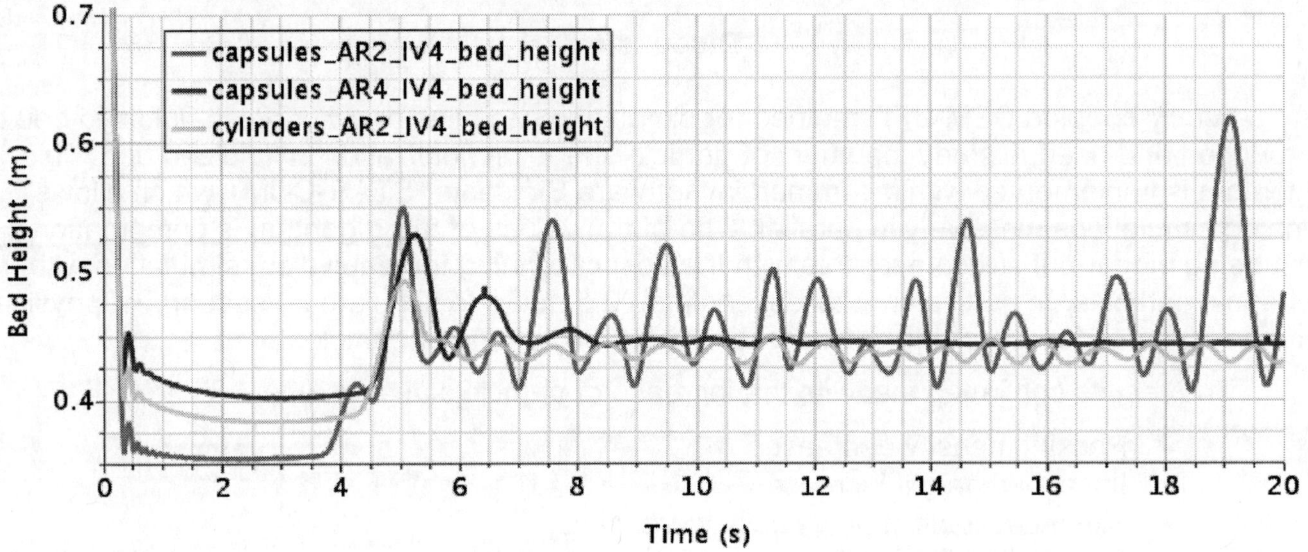

Figure 12. Pressure drop as a function of time

Bed height as a function of time is shown in Fig 12. It is consistent with previous results indicating more dense packing of short capsules compared to long capsules and short cylinders. It also shows much higher fluctuation in bed height in fluidized regime for short capsules compared to other two cases.

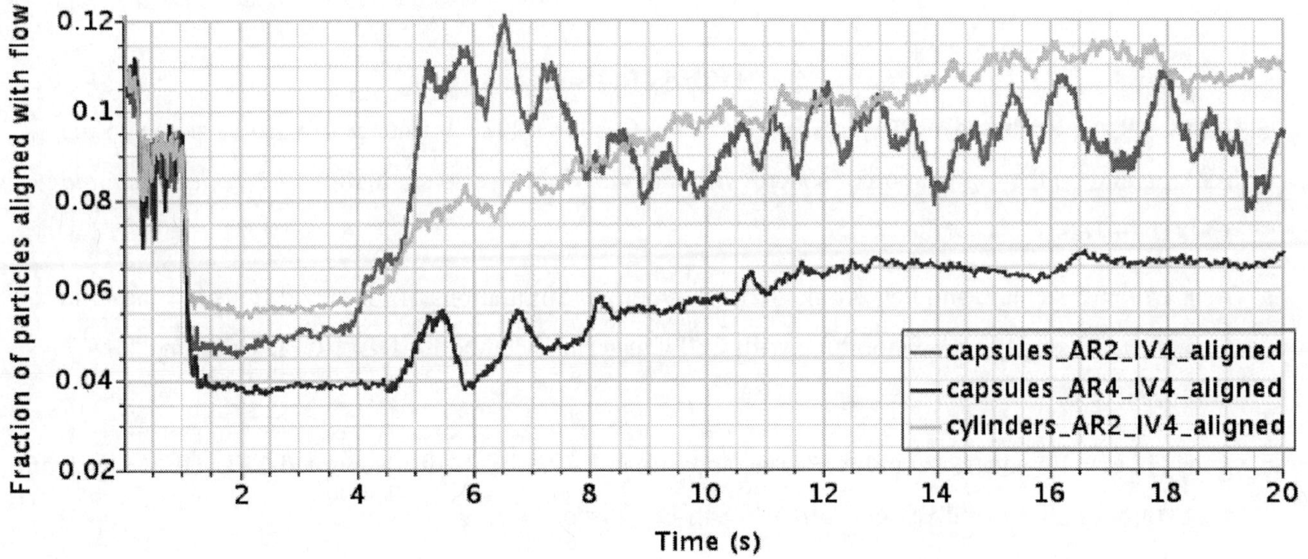

Figure 13. Ratio of aligned with the flow particles to all particles as a function of time

Finally, Fig. 13 shows the results for the number of aligned with the flow particles as a function of time. Unlike previous results for pressure, bed height, and void fraction, the steady state orientation of particle has much higher dependence on the aspect ratio of particles. In particular both short cylinders and short capsules end up with more particles aligned with flow direction in steady state compared to long capsules. These results will be discussed in more details next.

Discussion

A fully coupled DEM-CFD method for simulating capsules or cylinders in fluidized bed is introduced and used to study the effect of particle shape on fluidization in fluidized bed. This algorithm is implemented within commercial software Simcenter STAR-CCM+ which allows running simulations massively in parallel. One of the values of this algorithm is computational cost being bigger but still comparable with the cost of running the equivalent simulation with spherical particles. In particular, it takes less than 2.5 hours to simulate 20 seconds of physical time on 36 cores for any of the considered in this study shape of particles.

The results confirmed that with the increase of particle's aspect ratio

- Packing density decreases
- Initial bed height increases
- Minimum fluidization velocity increases
- Channeling flow regime is replacing bubbling flow regime for the same inlet velocity
- Takes longer to get steady state in particle's orientation

Similar changes but of lesser degree are observed when changing shape from capsule to cylinder while keeping the same aspect ratio. All recorded observation are consistent with the similar simulation or experimental results reported in the literature (Williams and Philipse 2003), (Buist, et al. 2017), (Cai, et al. 2016), (Mahajan, et al. 2017).

References

Abreu, Charlles R.A., Frederico W. Tavares, and Marcelo Castier. 2003. "Influence of particle shape on the packing and on the segregation of spherocylinders via Monte Carlo simulations." *Powder Technology* 134: 167– 180.

Baran, Oleh, Doran Greening, and Petr Kodl. 2016. "Numerical Simulations of Fluidized Beds in Fine-Grid Geometries Using Source Smoothing with Cell Clustering Method for DEM-CFD Coupling." *AIChE.* San-Francisco.

Baran, Oleh, Thomas Eppinger, Ravindra Aglave, and Simon Lo. 2016. "A fully coupled CFD-DEM algorithm for S-L mixing in stirred vessels." *AIChE.* San-Francisco.

Borzsonyi, Tamas, and Ralf Stannarius. 2013. "Granular materials composed of shape-anisotropic grains." *Soft Matter* 9: 7401.

Buist, Kay A, Pavithra Jayaprakash, J.A.M. Kuipers, Niels G. Deen, and Johan T. Padding. 2017. "Magnetic Particle Tracking for Nonspherical Particles in a Cylindrical Fluidized Bed." (AIChE Journal) 63: 5335.

Cai, Jie, Zhengbiao Peng, Charley Wu, Xiaobao Zhao, Zhulin Yuan, Behdad Moghtaderi, and Elham Doroodchi. 2016. "Numerical Study of the Orientation of Cylindrical Particles in a Circulating Fluidized Bed." *Ind. Eng. Chem. Res.* 55: 12806-12817.

Feng, Y.T., K. Han, and D.R.J. Owen. 2012. "Energy-conserving contact interaction models for arbitrarily shaped discrete elements." *Comput. Methods Appl. Mech. Engrg.* 205: 169-177.

Guo, Y., C. Wassgren, B. Hancock, and J. Curtis. 2013. "Granular shear flows of flat disks and elongated rods without and with friction." *Physics of Fluids* 25: 063304.

Liao, Lingzhi. 2013. "Influence of particle shape and bed height on fluidization." thesis, University of Florida.

Mahajan, Vinay, Tim M.J. Nijssen, Barry W. Fitzgerald, Jeroen Hofman, Hans Kuipers, and Johan T. Padding. 2017. "Fluidization of spherocylindrical particles." *Powders & Grains* 140: 06019.

Mema, Ivan, Vinay V Mahajan, Barry W Fitzgerald, Hans Kuipers, and Johan T Padding. 2017. "Effect of lift force on dense gas-fluidized beds of non-spherical particles." 12th International Conference on CFD in Oil & Gas, Metallurgical and Process Industries.

Williams, S. R., and A. P. Philipse. 2003. "Random packings of spheres and spherocylinders simulated by mechanical contraction." *Phys. Rev. E* 67: 051301.

Zhong, Wenqi, Yong Zhang, Baosheng Jin, and Mingyao Zhang. 2009. "Discrete Element Method Simulation of Cylinder-Shaped Particle Flow in a Gas-Solid Fluidized Bed." *Chem. Eng. Technol.* 32: 386-391.

APPLICATION OF CPFD METHOD IN THE SIMULATION OF VERTICAL DENSE PHASE PNEUMATIC CONVEYING OF PULVERIZED COAL

Yong Jin[†]*, East China University of Science and Technology, Shanghai, PRChina*
Haifeng Lu, East China University of Science and Technology, Shanghai, PRChina
Xiaolei Guo, East China University of Science and Technology, Shanghai, PRChina
Xin Gong[‡]*, East China University of Science and Technology, Shanghai, PRChina*

Abstract

Flow stability is a key issue to be addressed for the development of dense phase pneumatic conveying. Flow patterns are closely related with flow stability. From the view of experimental materials, angle of repose, packing density and tapped density were determined by PT-X powder tester. And flowability was determined by FT4 Powder Rheometer. The experimental results shew that the tested pulverized coal was a kind of Geldart's group C powder with strong cohesion. This study contributes with a computational particle fluid dynamic (CPFD) simulation which is based on the numerical solution of continuity and momentum balance equations in a three-dimensional (3-D) horizontal pipeline. A novel correction model was developed, describing interaction among particles and between fluid and particles decided by the experiment mentioned before. CPFD approach predicted the particle velocity and pressure drop were generally below 5% for the stable cases. What's more, the twelve different flow patterns were predicted by the CPFD approach embedded in the correction model, which was confirmed by the electrical capacitance tomography (ECT) in the experiment facility set up by our own. With the help of our correction model, it could reveal the mechanism of the change of flow patterns.

Keywords: CPFD; flow patterns; flow stability; ECT; correction model;dense

Highlights

Multiphase particle-in-cell model is used to predict dense phase pneumatic conveying.
Pressure drop and velocity profiles show good agreement with experiments.
A improved model was proposed to give the right simulation results.

1. Introduction

The practical application of the pneumatic conveying technology of particulate materials are extremely extensive, especially for the field of pulverized coal pressurization and gasification process. High pressure dense phase pneumatic conveying of pulverized coal is one of the key technologies of entrained flow gasification technology. The stability of transportation directly affects the gasification temperature, gasification products and the safety of the gasifier. Depending on the materials to be conveyed, operational conditions and facility geometries, transportation can

† Present author: Yong Jin, PhD student, E-mail: mhzxjinyong@126.com
‡ Corresponding author: Prof. Xin Gong, E-mail: gongxin@ecust.edu.cn

occur in different flow regimes, which usually can be categorized as dilute or dense. Despite wide applications, efficient pneumatic system design remains challenging and it is complicated by interactions between the particle phase and the turbulent carrier flow as well as particle-particle and particle-wall interactions. Because of these complications, a fundamental understanding of the complex physics, involved in gas-solids flows is crucial for the proper design, control and optimization of any pneumatic conveying systems.

For the dense phase pneumatic conveying system, transportation spped is low, the solid concentration is high and the flow patterns are complex, where the transportation requirements and conditions of the system are quite different from the general conveying system. Nowadays, many researches focused on the pressure drop analysis [1], flow patterns [2], signal analysis [3], where the research methods are experiment. However, the complex interactions among the particle phase and fluid phase are difficult for people to explore by experiments, like particle-particle and particle-fluid micro interactions, resulting the macro particle flow. Numerical simulation emerges at a historic moment. The computational particle fluid dynamics (CPFD) numerical scheme incorporated with the multiphase-particle-in-cell (MP-PIC) method is a recently developed model for gas-solids flows. To date, this method has been used to predict certain processes; however, to the authors' best knowledge, published information is unavailable on the application of this model in dense pneumatic conveying. Previous applications of the MP-PIC method include bubbling and circulating fluidized beds (CFBs), spouting beds, fluidized bed gasifiers (FBG), fluidized beds (FB) for carbon capture, gas/liquid/solid fluidized beds, Rayleigh-Taylor mixing layers and etc.

As in all other Eulerian-Lagrangian approaches, the MP-PIC method treats the gas as a continuous phase and the particles in a discrete manner. Solving a transport equation called a Liouville equation for the particle distribution function predicts the particle dynamics. The particle distribution function contains particle properties such as particle spatial location, particle velocity, particle mass, and time. The MP-PIC method uses the concept of parcels (numerical particles) in order to reduce the number of particles involved in the computations; this provides a significant acceleration of simulation speed. A parcel is a collection of several real particles with the same properties (species, size, density, temperature, etc.). Unlike DEM models that calculate particle-to-particle force with a spring–damper model and direct particle contact, the MP-PIC methodology models particle collision force for each particle as a spatial gradient. A particle normal stress model is developed from this concept to describe the particle collisions.

Therefore, this work uses the MP-PIC method to perform three-dimensional simulations of the dilute-phase pneumatic conveying of plastic pellets in a horizontal circular pipe.. In addition to investigating the validity of the method for pneumatic conveying applications, we also develop a model, describing interaction among particles and between fluid and particles decided by the experiment for the accurate prediction of pressure drops and flow patterns in dense phase pneumatic conveying. Moreover, some of the major model parameters deemed important in predicting pneumatic conveying characteristics, namely particle-particle and particle-wall collision parameters. The CPFD numerical methodology incorporated in the commercially available Barracuda® 15 code is the platform used for the modelling and simulations in this study.

2. Model description

2.1. Governing equations

The governing equations of the models for describing gas and solid phase are presented below. The gas phase equations consist of mass, momentum and energy conservation equation.

The gas phase continuity equation without interphase transfer is given by.

$$\frac{\partial \theta_f}{\partial t} + \nabla \cdot (\theta_f \mathbf{u}_f) = 0 \tag{1}$$

Where $\mathbf{u}_f$ is gas velocity and θ_f is gas volume fraction.

The gas phase velocity components in the *x*, *y* and *z* directions are given by time-averaged Navier-Stokes equations.

$$\frac{\partial(\theta_f \mathbf{u}_f)}{\partial t} + \nabla \cdot (\theta_f \mathbf{u}_f \mathbf{u}_f) = -\frac{1}{\rho_f}\nabla \mathbf{P} - \frac{1}{\rho_f}\mathbf{F} + \theta_f \mathbf{g} \tag{2}$$

Where ρ_f is fluid density, **P** is fluid pressure, **g** is gravity, **F** is the rate of momentum exchange per volume between the fluid and particles phases.

The constitutive equation for the non-hydrostatic part of the gas stress tensor is given by

$$\tau_g = \mu_{eff}(\nabla \mathbf{u}_g + \nabla \mathbf{u}_g^T) - \frac{2}{3}\mu_{eff}\nabla \cdot \mathbf{u}_g \mathbf{I} \tag{3}$$

where μ_{eff} is the shear viscosity, which is the sum of the laminar shear viscosity and the turbulent viscosity ($\mu_{t,g}$). The large eddy simulation (LES) model is used for calculating the gas phase turbulence. In LES, large eddies are calculated from the flow equations and the subgrid turbulence is captured with a Smagorinsky subgrid scale model.

$$\mu_{t,g} = C_S \rho_g \Delta^2 \left|\nabla \mathbf{u}_g + \nabla \mathbf{u}_g^T\right| \tag{4}$$

where Δ, $\Delta = (\delta x \delta y \delta z)^{1/3}$ is the subgrid length scale. Smagorinsky constant = 0.01 is used for the current simulations.

The interphase momentum transfer is given through an interphase viscous drag force ***F***.

Gidaspow [4,5] proposed a drag function by blending Wen-Yu and Ergun functions; this model is used as the drag function in the present study. A switching function is used in order to prevent possible numerical problems due to sharp transitions or discontinuities in the blended model. In the present study, $\theta_{cp} = 0.6$ is used.

$$D = \begin{cases} D_1 & \theta_p < 0.75\theta_{cp} \\ (D_2 - D_1)\left(\dfrac{\theta_p - 0.75\theta_{cp}}{0.85\theta_{cp} - 0.75\theta_{cp}}\right) + D_1 & 0.85\theta_{cp} \geqslant \theta_p \geqslant 0.75\theta_{cp} \\ D_2 & 0.85\theta_{cp} < \theta_p \end{cases} \tag{5}$$

Where, $D_1 = \frac{3}{8} C_d \frac{\rho_f |\mathbf{u}_f - \mathbf{u}_p|}{\rho_p r_p}$, $D_2 = 0.5\left(\frac{C_1 \theta_p}{\theta_f Re} + C_2\right)\frac{\rho_f |\mathbf{u}_f - \mathbf{u}_p|}{r_p \rho_p}$

The dynamics of the particle phase is predicted by solving a transport equation for the particle distribution function (PDF), $\varphi(\boldsymbol{x}, \mathbf{u}_p, m_p, \Omega_p, t)$ (Eq. (12)). Since the particles are non-vapourizing and chemically inert, the dependency of φ on particle temperature and material composition is ignored in this study.

The PDF $\varphi(\boldsymbol{x},\ \mathbf{u}_p,\ m_p,\ \Omega_p,\ t)$ is calculated from the Liouville equation

$$\frac{\partial \varphi}{\partial t}+\nabla(\varphi \mathbf{u}_p)+\nabla(\varphi A)=0 \tag{6}$$

Where $\boldsymbol{x}$ is the particle position, $\mathbf{u}_p$ is the particle velocity, m_p is the particle mass, Ω_p is the particle volume and t is the time

The trajectory of a given particle is solved by Newton's second law. The particle acceleration in the original MP-PIC method is

$$A = D_p(\mathbf{u}_f - \mathbf{u}_p) - \frac{1}{\rho_p}\nabla \mathbf{P} + \mathbf{g} - \frac{1}{\theta_p \rho_p}\nabla \tau \tag{7}$$

The particle volume fraction in the above equations is given by

$$\theta_p = \int\int\int \varphi \Omega_p \mathrm{d}\Omega_p \mathrm{d}\rho_p \mathrm{d}\mathbf{u}_p \tag{8}$$

Moreover, θ_p and θ_f are related by

$$\theta_p + \theta_f = 1 \tag{9}$$

τ in Eq. (7) is the particle stress function (also called particle normal stress or isotropic solid stress), which is used to model particle-particle interactions. Due to the difficulties in calculating the particle stress gradient for each particle in a dense flow, the particle-particle interaction forces are calculated as a spatial gradient (as a gradient on the grid) that is mapped back to the particles. The particle stress model used here is an extension of the model from Harris and Crighton [6].

$$\tau = \frac{P_s \theta_p^{\beta}}{\max[(\theta_{cp} - \theta_p), \varepsilon(1-\theta_p)]} \tag{10}$$

where P_s is constant with units of pressure, β is supposed to lie between 2 and 5 [7], and ε is a small number on the order of 10^{-7} to remove singularity. The values used in the current simulations are 1, 3 and 10^{-8}, respectively. However, the particle stress model has a negligible effect on the particle behaviour in the dense phase. The stress model will begin to affect particles as the particle volume fraction of a region begins to approach the close-packed level.

2.2. Model modification

For the particle force term, the external force term was modified by the the polydisperse system contact model of Liu et al [8], which was based on the Rumpf [9] equation, a model with elastic-plastic deformation was considered for the granular system:

$$\sigma_{t,iso} = \frac{1-\varepsilon}{\pi}\sum_{j=1}^{n} f_j d_j \sum_{i=1}^{n} k_j F_{ji} \Big/ \sum_{j=1}^{n} f_j d_j^3 \tag{11}$$

Here, the interaction force between particle i and particle j is F_{ji} :

$$F_{ji} = F_{H0}(j,i) + h(R_{ji}^{*})[F_{H0}(j,i) + F_N] \tag{12}$$

3 Computational details

The geometry, particle properties and operating conditions of the present simulations are based on self-established experimental facilities. work done by Tsuji and Morikawa [64]. These details including solution procedure are explained in the Subsections; 3.1–3.6. The fluid flow is isothermal, compressible and no reactions are involved. The effect of gravity is included.

3.1. Geometry and grid generation

The closed-loop experimental system designed to investigate the dense phase flow of pulverized coal is illustrated in Figure 1. The mass flow rates of three gaseous streams feeding the conveying system were individually controlled to achieve the desired operating conditions. The ECT sensor was installed on the 50 mm I.D. vertical pipe to monitor flow patterns. Two pressure transducers were mounted at the top of the horizontal pipe to obtain the wall static pressure signals at a sampling frequency of 1 Hz for a total of more than 200 sample points. The ECT and pressure signals were both recorded in a computer data acquisition system. The test section with the ECT sensor was located near the pressure transducers. Dry compressed air was used as the conveying gas.

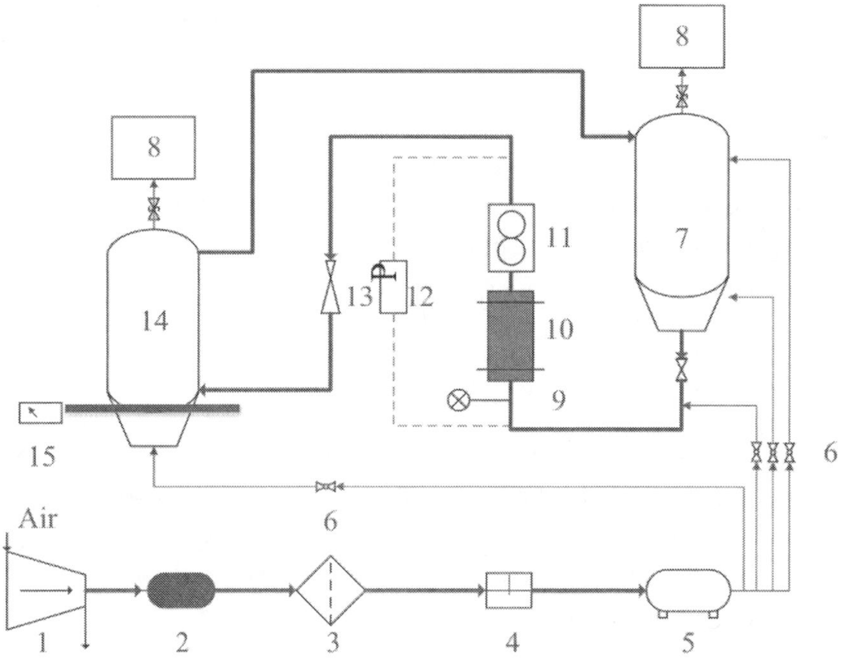

Figure 1. Schematic diagram of the vertical experimental facility.

1-air compressor; 2-buffer tank; 3-refrigeration compressed air drier; 4-filter; 5-gas distributor; 6-gas flowmeter; 7-feeding tank; 8-fabric filter; 9-pressure sensor; 10-solid mass flowmeter; 11-ECT sensor; 12-differential pressure sensor; 13-venturi; 14-receving tank; 15-weighing module.

A three-dimensional 4.2 m vertical circular pipe with an internal diameter of 50 mm was created (Figure 2 (left)). The geometry was meshed by a built-in module in Barracuda software. The mesh of the pipe cross section is shown in Figure 2 (right). The mesh had a total number of 89000 cells (5 × 5 × 5) and these were cube. The computational time was approximately 1–3 days for a high performance computer (Intel® Xeon® Processor X5650) characterized as 12 M Cache, 2.66 GHz, RAM 24 GB and 2 TB.

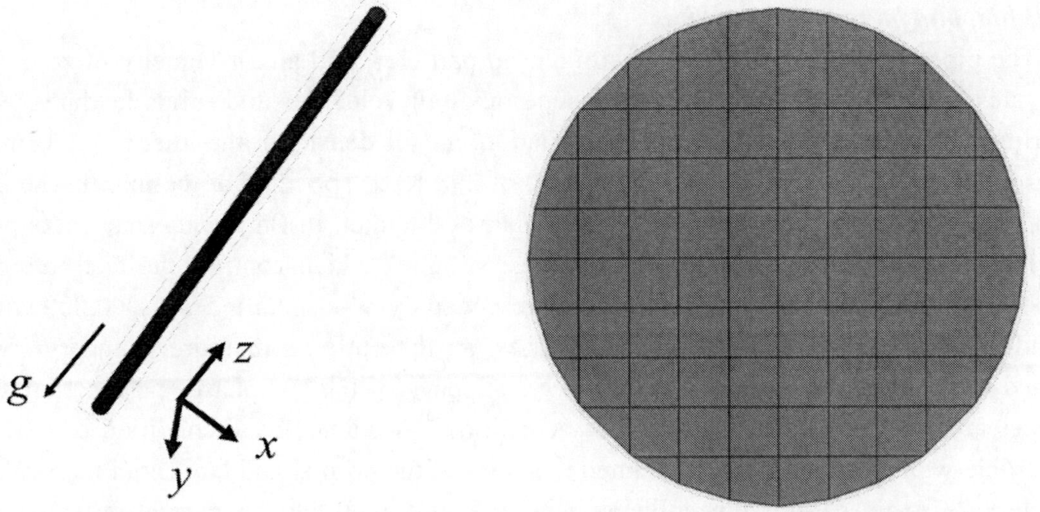

Figure 2. The computational domain (left) and grid of pipe cross section (right).

3.2. Material properties

The particles were lignite pulverized coal with a particle density of 1530 kg·m^{-3}. Particles were considered to be spherical. The used particle size distribution and characteristic particle size are shown in Fig. 2. The air viscosity was 1.84469 × 10^{-5} kg·m^{-3}.

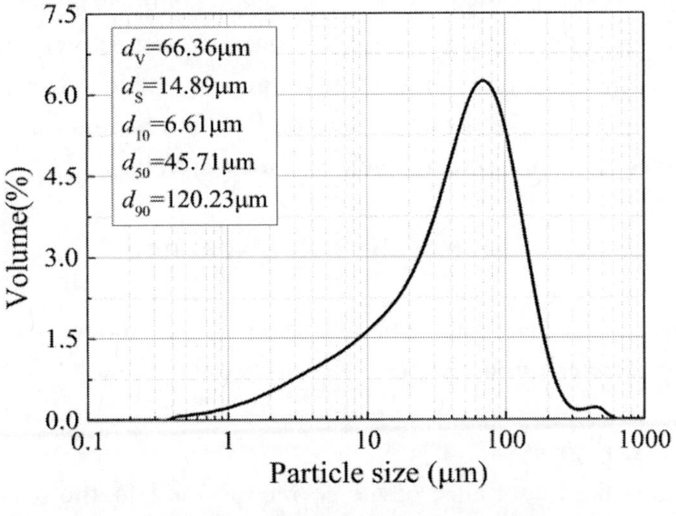

Figure 3. Particle size distribution of plastic pellets.

Table 1 Material properties of pulverized coal

ω(%)	θ_r(°)	θ_f(°)	θ_d(°)	θ_s(°)	ρ_b(g·cm^{-3})	ρ_t(g·cm^{-3})	ρ_p(g·cm^{-3})
3.35	50.09	28.09	22.00	60.70	0.4475	0.8593	1.490

According to the Carr's flowability index chart[10], the flowability index of our tested pulverized coal is 37, between 20 and 39, which are classified as the poor flowability powder, like Geldart's Group C powder. [11] Usually, this kind of powder shows great cohesive and there is a necessity for people to use detailed force equations to describe the motion of particles. However, in an MP-PIC scheme, each cell contains a number of numerical particles (parcels). Particle properties are mapped to and from the Eulerian grid to obtain grid-based properties for the particles; fluid properties are, in turn, mapped to discrete particle locations.

3.3. Initial and boundary conditions

The pipe is initially filled with 1 atm air, no particles, and an air velocity of zero. The mass flow rates are calculated according to the superficial air velocities and solids loadings used in the experiments. As for the mass flow rate calculation, the air density at the inlet is 1.2 kg·m^{-3}, which corresponds to 1 atm pressure at a temperature of 298 K, as specified at the inlet in the Barracuda setup. The solids have the same velocity to the air at the inlet. In Barracuda setup, it is possible to specify "the numerical particle feed per average volume" which controls the feed rate of parcels (numerical particles) into the domain. In the present study, it is specified as 1,000,000 which gives around 2–4 real particles/parcel for all the cases. At the outlet, a pressure boundary condition is adopted with a specified pressure of 1 atm. No slippage of the gas phase and a partial slip of the particles are assumed as the wall boundary conditions. A bounce-back condition is used to model the particle-wall interactions. It is assumed that 95% of the normal and tangential momentum of the particles are retained after the collision with the pipe wall, i.e., a normal-to-wall momentum retention = 0.95 and a tangent-to-wall momentum retention = 0.95 in the Barracuda setup.

3.4. Solver numeric and convergence criterion

The equations are solved in a transient state. An initial time step of 0.0001 s is applied to the CPFD simulation. However, there is a built-in-model in Barracuda that uses a Courant-Friedrichs-Lewy (CFL) algorithm to automatically determine the proper time step based on the simulated meshes and parcels in order to maintain stability, accuracy and calculation speed [13]. The lower and upper limits for CFL are 0.8 and 1.5, respectively. Each simulation was run over 6 s (in real time) to achieve a quasi-steady state. No major changes were observed in the variables after 1 s. As for the level of accuracy, the limited residual values for convergence were 10^{-7}, 10^{-8} and 10^{-7} for volume, pressure and velocity, respectively.

4. Results and discussion

The simulations are carried out with a certain set of model parameters, which will be published after the acceptance of our paper.

4.2. Prediction of pressure drop

Having determined the importance of the correction models, the results from the simulations carried out with the new collision terms are compared with experimental data.

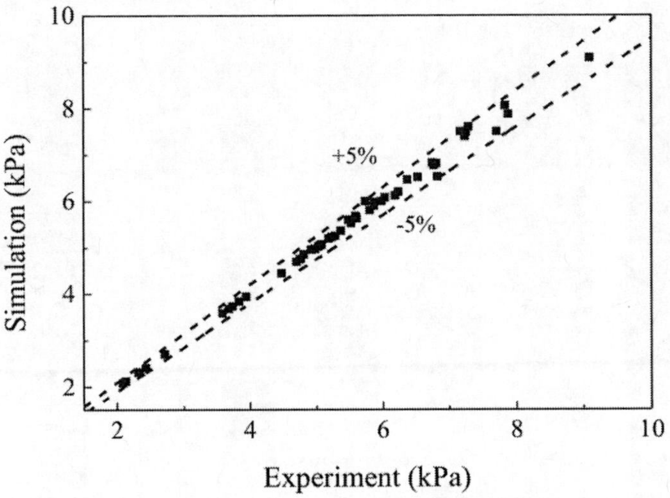

Figure 4. Comparison of 52 groups pf experimental data and simulation results

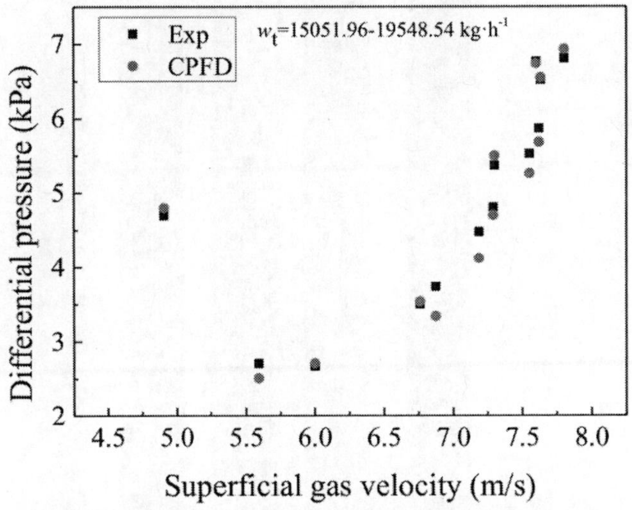

Figure 5. Phase diagram simulation (14 groups)

As shown in Figure 4, the predicted pressure drop results that the model provide better agreement with the experimental results. The predictions deviate from the experimental data is approximately 20%. Figure 5 gives the simulation of phase diagram.

4.2. Prediction of particle velocity

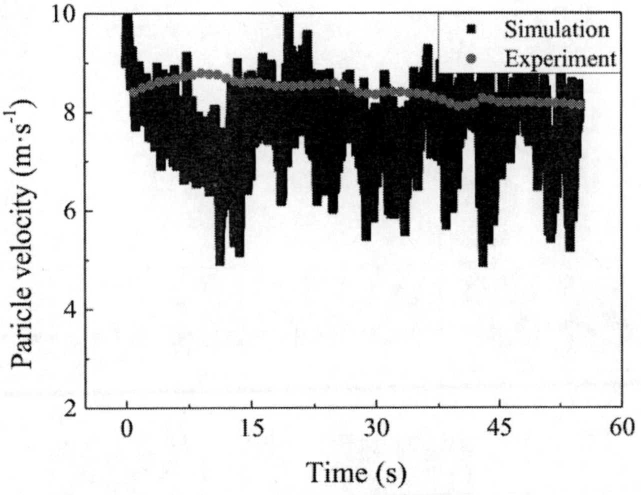

Figure 6. Comparison of simulated particle velocity and experiment velocity

Figure 6 shows that the deviation is about 10%.

4.3. Prediction of flow patterns

Table 2 give the simulation results and experiment results which shows great agreement.

Table 2 The relationship between superficial gas velocity and different flow patterns.

a	b	c	d	e	f	g	h	i	j	k	l

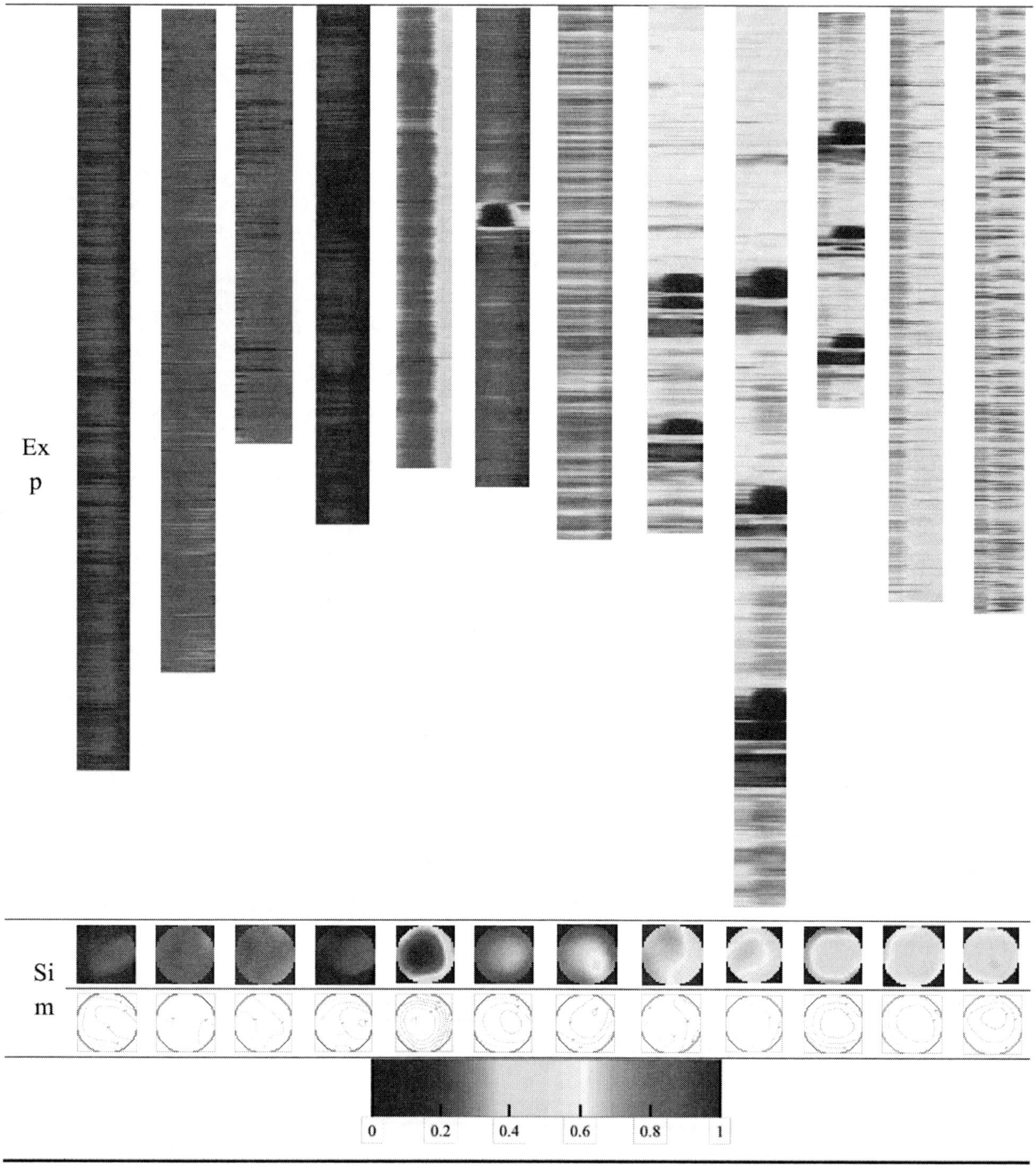

5. Conclusions

In this paper, a novel correction model was proposed to describe interaction among particles and between fluid and particles decided by the experiment mentioned before. CPFD approach predicted the particle velocity and pressure drop were generally below 5% for the stable cases. What's more, the twelve different flow patterns were predicted by the CPFD approach embedded in the correction model, which was confirmed by the electrical capacitance tomography (ECT) in the experiment facility set up by our own. With the help of our correction model, it may help people understand the change of flow patterns.

Acknowledgements

This work was supported by the National Natural Science Foundation of China (21206041), and Shanghai Science and Technology Commission Project (14DZ1200100, 15DZ1200802), the Fundamental Research Funds for the Central Universities (222201717004, 222201514338) and the key scientific research project in Shanxi (MH2014-01).

References

1. CW A, JH H, Hp S. Pressure Drop in Flow of Dense Coal-Air Mixtures. *Ind.Eng.Chem.* 1951;43(8):1837-1840.

2. Cong X, Guo X, Lu H, et al. Flow patterns of pulverized coal pneumatic conveying and time-series analysis of pressure fluctuations. *Chemical Engineering Science.* 2013;101:303-314.

3. Xingliang Cong. Study on the relationship between the flow type and the pressure signal in the pipeline. Shanghai, East China University of Science and Technology, 2012. (doctoral dissertation)

4. Wen CY, Yu YH. Mechanics of fluidization. *Chem.Eng.Progr.Symp.* 1966;62:100-110.

5. D.Gidaspow. *Multiphase Flow and Fluidization, Continuum and Kinetic Theory Description.* Boston: Academic Press; 1994.

6. Harris SE, Crighton DG. Solitons, solitary waves, and voidage disturbances in gas-fluidized beds. *J. Fluid Mech.* 1994;266:243 - 276.

7. Auzerals FM, Jackson R, Russel WB. The resolution of shocks and the effects of compressible sediments in transient settling. *Journal Fluid Mesh.* 1988;195:437-462.

8. Liu Y, Lu H, Poletto M, Guo X, Gong X. Bulk flow properties of pulverised coal systems and the relationship between inter-particle forces and particle contacts. *Powder Technology.* 2017.

9. Rumpf DIH. Grundlagen und Methoden des Granulierens. *Chemie Ingenieur Technik.* 1958;30(3):144-158.

10. Carr.R.L, Div.B. Evaluating flow properties of solids. *Chemical Engineering.* 1965;18(1):163-168.

11. D.Deldart. Types of gas fluidization. *Powder Technology.* 1973;7:285-292.

509860 Electrostatics of Dry Powder Aerosols for Inhalation

Abstract:

Electrostatics of pharmaceutical aerosols for inhalation is a relatively new research area. Particle charges have been proposed to potentially influence deposition in the lungs but no conclusive in vivo data have been reported. Investigation of the relationship between the formulation and aerosol charging is required to further explore this area. The modified electrical low pressure impactor (ELPI) and bipolar charge analyser (BOLAR) have been used to measure the size and electrostatic charge of particles generated from dry powder inhalers (DPIs). DPIs are commonly used to deliver drugs to the lungs. The powder is dispersed by the inhalation effort of the patient. Net charges are generated through triboelectrication during aerosolisation. These charges were found to be formulation- and humidity-dependent [1, 2], as well as bipolar [3, 4]. The knowledge gained may be useful in the development of pharmaceutical aerosols and pulmonary drug delivery in the future.

References

1. Philip Chi Lip Kwok, Hak-Kim Chan (2008) Effect of relative humidity on the electrostatic charge properties of dry powder inhaler aerosols. Pharmaceutical Research, 25(2): 277-288.
2. Jennifer Wong, Philip Chi Lip Kwok, Tim Noakes, Ali Fathi, Fariba Dehghani, Hak-Kim Chan (2014) Effect of crystallinity on electrostatic charging in dry powder inhaler formulations. Pharmaceutical Research 31(7): 1656-1664.
3. Jennifer Wong, Yu-Wei Lin, Philip Chi Lip Kwok, Ville NiemelÃ¤, John Crapper, Hak-Kim Chan (2015) Measuring bipolar charge and mass distributions of powder aerosols by a novel tool (BOLAR). Molecular Pharmaceutics 12(9): 3433-3440.
4. Jennifer Wong, Philip Chi Lip Kwok, Ville NiemelÃ¤, Desmond Heng, John Crapper, Hak-Kim Chan (2016) Bipolar electrostatic charge and mass distributions of powder aerosols â€“ Effects of inhaler design and inhaler material. Journal of Aerosol Science 95: 104-117.

Copyright © American Institute of Chemical Engineers. All rights reserved.

507600 Electrification and Dispersion of Particles Using Mesh Electrode

Abstract:

Electrostatic forces cause spontaneous movement in every charged particle. Thus, electrostatic technologies are attracting attention in the actual processes of powder handling, i.e., separation, classification, and collection. When conductive particles come in contact with any one of the inside walls of parallel electrodes, the particles are charged to the same polarity as that of the electrode by induction. If Coulomb forces acting on the charged particles are larger than the adhesion forces between the contact surfaces, the particles will levitate from the electrode. After the particles move and reach the counter electrode, the polarity of the particles inverts, and then the particles levitate again. As a result, the particles oscillate between the parallel electrodes. Replacing the counter electrode with a mesh electrode allows the passage of the charged particles through the mesh openings; then, the particles with the same polarity as the first electrode are dispersed by mutual electrostatic repulsion. However, since the particles are polarized in the external electric field, chain agglomerates can be formed; thereby the particle behavior becomes more complicated. In powder handling processes, dielectric particles are often used. Such particles are accumulated and form multi-layers on the electrodes; this affects the phenomena of particle charging and levitation. The phenomena for single dielectric particles have been reported, but there are few studies focusing on the induction charging and levitation for particle layers.

We present the charging, levitation, and relevant behaviors for dielectric particle layers, which are obtained using parallel electrodes consisting of a lower plate electrode and an upper mesh electrode. First, multi-layers of particles are formed on the lower electrode. Next, DC voltage is applied to the lower electrode and the upper electrode is grounded to vertically form a high-intensity electric field. Then, the particles are charged and levitated in the electric field. The particle motion around the lower electrode is observed via a high-speed microscope camera, showing that agglomerated as well as single particles levitate from the surface of the particle layers, i.e., the particles are polarized in the electric field and form chain agglomerates by mutual electrostatic interactions. When the specific charge, i.e., charge-to-mass ratio of particles is small, single particles cannot levitate because of small Coulomb forces. However, chain agglomerates can levitate because of large Coulomb forces even though the specific charge is small. The Coulomb forces of the chain agglomerates can exceed the adhesion forces and gravitational forces acting on primary particles in the agglomerates.

The specific charge of each levitating particle is estimated by fitting experimentally obtained particle trajectories with calculated ones in simulated electric fields, showing that as the number of primary particles of the chain agglomerate increases, their specific charge decreases. In addition, it is found that as the adhesion force of each primary particle increases, the number of primary particles of the chain agglomerate increases.

Furthermore, a unique phenomenon is observed around the upper electrode. Fig. 1 shows a chain agglomerate that was dispersed. Agglomerates can be rotated in the dispersion process. The charges of the agglomerates during the dispersion process are estimated using a simulated non-uniform electric field around the upper electrode and the effect of the particle charge on the dispersion process is studied.

Fig. 1. Dispersion of chain agglomerate by electrostatic repulsion around upper mesh electrode.

Copyright © American Institute of Chemical Engineers. All rights reserved.

NOVEL ELECTROSTATIC FIELD METER USING ROLLING SECTOR

(FIRST PAPER)

Dr. Kwangseok Choi, National Institute of Occupational Safety and Health, Japan, Tokyo, Japan
Dr. Teruo Suzuki, Kasuga Denki Inc., Kanagawa, Japan

Abstract

This paper investigated experimentally the fundamental characteristics of the novel electrostatic field meter in order to measure electrostatic charge levels. The novel meter consists mainly of a grounded plate with a sensing rolling sector, the electric motor, a signal cable, an enclosure, and two air supply lines. The grounded plate had four 0.9 mm holes that had a trench shape to satisfy the international standard. The rolling sector was rotating at 1100 rpm. The enclosure was supplied with compressed air (air purge) to prevent combustible gases or powders from getting inside. As the experimental demonstrated, (1) the electrostatic field strength of the simulated charge material and detected signal were clearly in direct correlation, and its validity was confirmed; (2) the signal of the novel meter strongly depends on the thickness of the grounded metal plate; and (3) the performance of the novel meter was not affected by air purge in this study.

Keywords: *electrostatic charges, electrostatic discharges, electrostatic field, powder, explosion*

1. Introduction

Electrostatic charges and discharges generated in industry can cause gas and dust explosions and damage electronic components. Especially, unwanted electrostatic charges and discharges can cause major damage such as fires in buildings, damage to sensitive electrical equipment, or injury to humans, including death. Many papers have reported major accidents caused by electrostatics in the industrial process [1–4].

This work deals with the experimental investigation of a novel electrostatic field meter (hereafter, the novel meter) based on its performance on a laboratory scale as a first paper.

2. Experimental apparatus and method

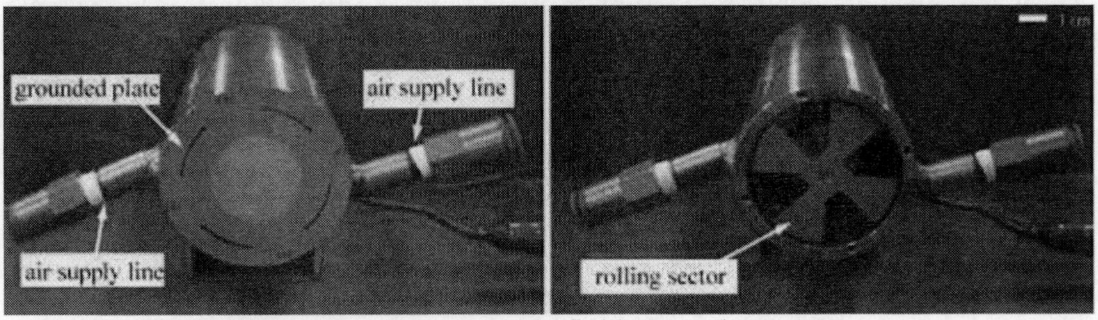

Fig. 1 Novel electrostatic field strength sensor

Figure 1 shows that the novel meter consists mainly of a grounded plate with a sensing rolling sector, electric motor, signal cable, enclosure, and two air supply lines. The grounded plate was attached to the front of the enclosure. It had four 0.9 mm holes that had a trench shape to satisfy the requirement of the International Electrotechnical Commission (IEC)-Ingress Protection (IP) code [5]. The sensing rolling sector (signal-detecting electrode) was placed 1 mm behind the grounded plate. Using the electric motor, the signal-detecting electrode rotated at 1100 rpm. To measure signals, the rolling sector was separated from the motor's axis by an insulator. The enclosure was supplied with compressed air (air purge) to prevent combustible gases or powders from getting inside. In this study, a laboratory test was conducted to assess the performance of the novel meter. The experimental setup consisted of the novel meter, a rotating sector controller, an HV power source, a charged metal plate (500 mm × 500 mm in area, 2 mm thick), a digital oscilloscope, an air compressor, an automatic air dryer, an electrometer, a mass flow meter, and other auxiliary devices. The charged metal plate in this study was assumed to be the charging material and was attached to an HV DC power source to control its charge level. The novel meter was fixed toward the center of the charged metal plate. The novel sensor is a device based on electrostatic induction. A 100 M□ was installed inside the electric circuit to convert into voltage in order to measure the signal. The effects of the electrostatic field E [kV/cm], the thickness t [mm] of the grounded plate, and the air purge P [MPa] on signal V [mV] were examined in this study.

3. Experimental results and discussion

Figure 2 shows the effect of the voltage applied to the charged metal plate on the signal detected by the novel meter without purging air. The applied voltage means the potential difference between the novel meter and the charged metal plate. Because the distance between the novel meter and the charged metal plate was 10 mm in this study, the values of the applied voltage and the electrostatic field were the same. The range of applied voltage to the metal plate was in 0 to10 kV in bipolar. It should be noted here that the signal of the novel meter was almost zero when high voltage was not applied to the metal plate. As for the results, the applied voltage had a positive correlation to the detection signal. The positive and negative values were very similar, only with different signs, showing a fairly linear graph when plotted. This shows that, as voltage increased, the detected signal increased as well. That is to say, electrostatic field E of the simulated charging material (metal plate) and detected signal V were clearly in direct correlation, and its validity was confirmed.

Figure 3 shows the effect of the thickness, t, of the grounded metal plate on the signal detected by the novel meter when V and d were 10 kV and 10 mm, respectively, also without the air purge. The thickness was in a range of 0.5 mm to 2.0 mm, every 0.5 mm. Results showed that thickness had a negative effect on the detection signal. For example, detected signals V for t : 0.5 mm, 1.0 mm, and 1.5 mm were 26.6 m, 5.87 mV, and 1.53 mV, respectively. When the thickness was 2.0 mm, the detected signal was almost zero. This can be explained by the fact that the electric field generated by the charged metal plate could not fully reach the rotating sector because it headed toward the grounded metal plate.

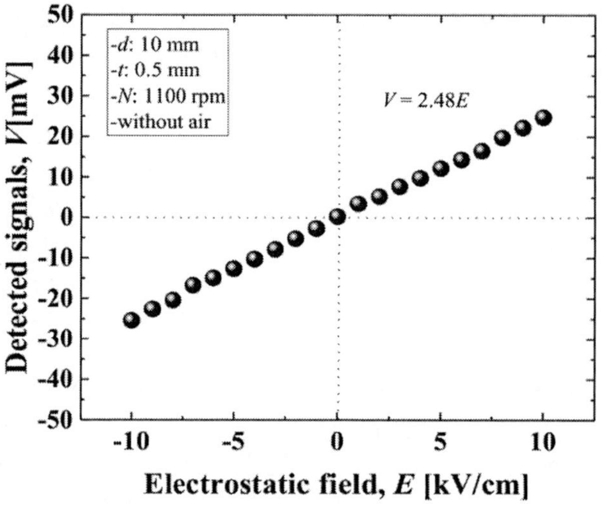

Fig. 2 Relationship between the electrostatic field and the signal of the novel meter

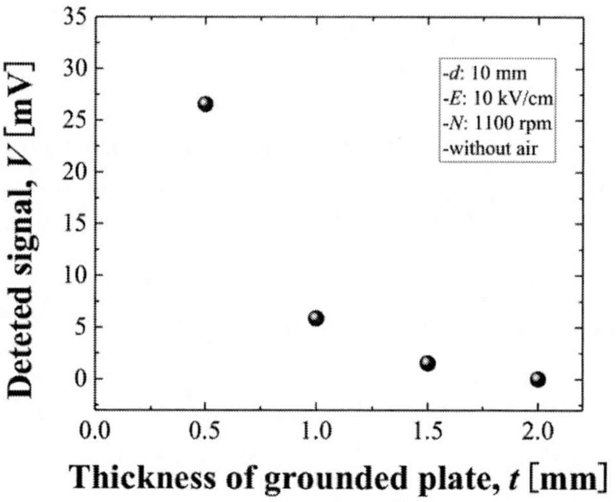

Fig. 3 The signal of the novel meter as a function of the thickness of the grounded metal

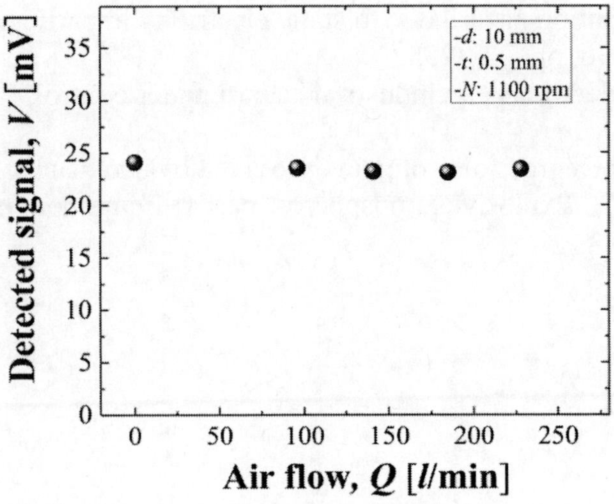

Fig. 4 Relationship between the detected signal and the air purge of the novel meter

Figure 4 shows the effect of the air purge on the signal detected by the novel meter when E was10 kV. The flow of air for the purge, Q, in the novel meter ranged between 0 and 230 l/min. The pressure used for purging in the novel meter was also measured. Results show that the performance of the novel meter was almost unaffected by the purge of air supplied to protect combustible gases or powders from getting inside. It is notable that even in processes with combustible materials, the novel meter can still give an accurate signal. The pressure in the novel meter due to the purge of air increased as the air flow increased. When the air flow was 96 l/min, 141 l/min, 185 l/min, or 230 l/min, the pressure was 212 Pa, 443 Pa, 734 Pa, and 1020 Pa, respectively. Air flow of the novel meter at or above 96 l/min can satisfy the requirement for an internal pressure considered explosion-proof according to the IEC standard, which is 50 Pa [6].

4. Conclusion

This paper investigated experimentally the effect of the electrostatic field, the thickness of the grounded plate, and the air purge pressure on its performance in a laboratory setting. A summary of the experimental results follows:

1) The tests confirmed the linearity of the relationship between the detected signal and the electrostatic field.

2) The detection signal decreased as the thickness of the grounded plate increased. When the thickness was 2.0 mm, the detected signal was almost zero.

3) The air flow (max. 230 l/min.) for purging the novel meter did not have an effect on the detected signals.

References

1. T. H.Pratt (2000): Electrostatic ignitions of fires and explosions, Center for chemial process safety, pp.132-142.
2. F. Riedewald (2012): A fire in a secondary pharmaceutical powder transfer operation, Process safety progress, 31, 4, pp. 390-392.

3. M. Glor (2003): Ignition hazard due to static electricity in particulate processes, Powder technology, 135, 136, pp. 223-233.
4. N. Gibson: Static electricity- an industrial hazard under control, J. electrostatics 40, 41, pp.21-30.
5. IEC 60529 (2004), Degrees of protection provided by enclosures (IP code).
6. IEC 60079-2 (2015), Explosive atmospheres- part 1: Equipment protection by flame proof enclosures.

507317 Investigation of Granular Surface Roughness Effect on Electrostatic Charge Generation

Abstract:

Abstractï¼šElectrostatic charge generation is a multivariable and complex process whose working mechanism has never been fully understood. The objective of this study is to investigate the effect of granule surface roughness on electrostatic charge generation. Two kinds of granule material, namely Polyvinyl chloride (PVC) and polypropylene (PP) were used with the granule size of 4mm in diameter and 2mm in height and in the shape of either cylinder or semi-cylinder. The working surfaces were grounded and the corresponding roughness ranged from 0.140 to 8.600 Î¼m. Uneven surfaces were found to give rise to voids between two solids, where air stored in the voids was able to accelerate discharging. With the same roughness, PVC tended to generate more electrostatic charge than PP by one order of magnitude. For both materials, electrostatic charge generation first increased with surface roughness and then decreased. The maximum electrostatic charge generation was found to occur when the effects of interaction, contact area and voids discharging were at equilibrium. With the combined effect of humidity, surface roughness and contact area, highest electrostatics generation occurred near the mid-roughness point tested in this work. Humidity had more effect on electrostatic charge generation as the granule working surface had lower roughness.

Keywords: Electrostatic charge generation; Surface roughness, Granule.

Copyright © American Institute of Chemical Engineers. All rights reserved.

504580 Contact Charging in Granular Materials

Abstract:

Contact charging of fine, sub-millimeter particles and the resulting clustering is important in circumstances ranging from the early stages of planet formation to industrial powders to airborne pollutants. Even in systems comprised of grains of identical dielectric material, contact charging can generate large amounts of net positive or negative charge on individual particles, resulting in long-range electrostatic forces. Remarkably, even basic aspects of contact charging, such as the nature of the charge carriers or the charge transfer mechanism are still under debate. This talk focuses on recent work where collision events between individual particles are tracked with high-speed video and the charge on single particles can be extracted. In freely falling granular streams we observe collide-and-capture events between charged particles and particle-by-particle aggregation into clusters. Size-dependent contact charging is found to produce a variety of charge-stabilized â€œgranular moleculesâ€, whose configurations can be modeled by taking many-body dielectric polarization effects into account. I will also introduce a new approach, based on ultrasonic levitation, for studying contact charging of single particles. This method allows for measurements under a wide range of environmental conditions as well as applying an electric field, and its exquisite sensitivity makes it possible to determine the charge transferred in a single collision.

Copyright © American Institute of Chemical Engineers. All rights reserved.

502848 Mesoscience: Opening a New Paradigm of Particle Technology

Abstract:

Mesoscience: ***Opening a new paradigm of particle technology***

Jinghai Li

National Key Laboratory of Multi-phase complex systems

Institute of Process Engineering, Chinese Academy of Sciences

Beijing 100190, P. R. China

Email: jhli@ipe.ac.cn

Mesoscale phenomena exist in between "unit" scales and "system" scales at different levels of the real world, spaning from elemental particle to the universe, as indicated in the following Figure[1]. Understanding of complex processes at mesoscales, characterized by spatiotemporal dynamic structures, is a common challenge for the whole spectrum of science and technology.

This presentation reviews the three decades of research on mesoscales of particulate systems at IPE-CAS. It was initiated by the energy-minimization multiscale (EMMS) model specific for gas-solid fluidization[2], which established the stability condition or variational function for particle clustering phenomenon. In extending the model to many different complex systems, such as gas-liquid, turbulence, gas-liquid-solid, emulsion, material preparation, protein, and catalysis systems, the EMMS principle was then proposed generally for different complex systems, featuring compromise in competition between different dominant mechanisms in physics and formulated as multi-objective variational problem in mathematics[3]. With increasing understanding of the generality of this principle, the concept of mesoscience was further advanced[4, 5], and is believed to be potentially universal for all complex mesoscale phenomena at different levels.

The presentation will also give a perspective on mesoscience. It is believed that more evidence will be needed from various disciplines[4, 5], particularly, from the field of particle technology where mesoscale phenomena occur everywhere such as in formulating and processing particles. The development of mesoscience will enable our capability in particle design, rational synthesis, smart massive production and system optimization due to its underlying principle to bridge unit scales and system scales. In contributing more evidence to mesoscience in exploring its universality, in return, the knowledge, tools and methods in particle technology will be upgraded to a new paradigm[6]. In such a paradigm, the theory, computation and experiment at mesosclaes will be dominant, and the virtual process engineering will be enabling[7, 8].

References

1. J. Li, W. Ge W. Wang, N. Yang, W. Huang, Focusing on mesoscales: from the energy-minimization multiscale model to mesoscience, Current Opinion in Chemical Engineering, 2016, 13: 10-23.

2. J. Li, Y. Tung, M. Kwauk, Method of energy minimization in multi-scale modeling of particle fluid two-phase flow, *Circulating Fluidized Bed Technology II*, Basu and J. F. Large (eds), Pergamon Press, 1988, 89-103.

3. J. Li, J. Zhang, W. Ge, X. Liu, Multi-scale methodology for complex systems, *Chemical Engineering Science*, 2004, 59(8-9): 1687-1700.

4. J. Li, W. Ge, W. Wang, N. Yang, X. Liu, L. Wang, X. He, X. Wang, J. Wang, M. Kwauk, *From multiscale modeling to meso-science -- a chemical engineering perspective*, Springer, Berlin, 2013.

5. J. Li, W. Huang, *Towards mesoscience: the principle of compromise in competition*, Springer, 2014.

6. J. Li, Exploring the logic and landscape of the knowledge system: multilevel structures, each multiscaled with complexity at the mesoscale, Engineering, 2016, 2: 276-285.

7. W. Ge, W. Wang, N. Yang, J. Li, M. Kwauk, F. Chen, J. Chen, X. Fang, L. Guo, X. He, X. Liu, Y. Liu, B. Lu, J. Wang, J. Wang, L. Wang, X. Wang, Q. Xiong, M. Xu, L. Deng, Y. Han, C. Hou, L. Hua, W. Huang, B. Li, C. Li, F. Li, Y. Ren, J. Xu, N. Zhang, Y. Zhang, G. Zhou, G. Zhou, Meso-scale oriented simulation towards virtual process engineering (VPE) -- the EMMS paradigm, Chemical Engineering Science, 2011, 66: 4426-4458.

8. X. Liu, L. Guo, Z. Xia, B. Lu, M. Zhao, F. Meng, Z. Li, J. Li, Harnessing the power of virtual reality, Chemical Engineering Progress, 2012, 108: 28-33.

Copyright © American Institute of Chemical Engineers. All rights reserved.

504481 Nature-Inspired Chemical Engineering - a Pathway to Innovation in Particle Technology

Abstract:

From the way trees grow and what makes our lungs so efficient, to what renders bacterial communities resilient or how regular patterns form in the sand by the action of the wind â€" nature holds a treasure trove of ideas to inspire solutions to technological problems. These include some of our most challenging problems in manufacturing, energy, sustainability or healthcare. Many of these problems relate to particle technology. Most often, nature's best solutions to a problem extend beyond first appearances and superficial similarities, and, so, observing nature should go beyond mimicking. Gaining inspiration from nature is most effective in solving technical problems when we have sufficient fundamental understanding that can then be appropriately translated within the context of an application. As in any product and process design, the real power of nature-inspired design requires moving beyond biomimicry, and appreciating the technical, industrial or societal context.

In my presentation, I will illustrate the thematic, mechanistic approach underpinning «nature-inspired chemical engineering» (NICE) and then apply it to timely problems in particle technology. I will focus on three key mechanisms that are ubiquitous in nature. First, efficient hierarchical transport networks, which allow for optimal scalability. Second, balancing of various forces, on large scales (mechanics) but also at the nano-scale, leading to confinement effects, where electrostatics play a key role in issues around selectivity and stability, as well as activity for catalysis or permeation for membranes. Third, dynamic self-organization, which is key to resilience and self-healing properties, as well as pattern formation, both in living and non-living systems.

For applications, we will consider problems in gas-solid fluidization, as well as in the design of hierarchically structured particles, which combine nano-confinement effects and optimized transport across length scales. The NICE approach leads to unexpected, out-of-the-box solutions (innovation), but also to new fundamental insights in, for example, fluidization, especially to tackle outstanding questions in particle technology that revolve around mesoscopic physics, which are key for engineers to translate science at the microscopic scale to the macroscopic world of applications.

Copyright © American Institute of Chemical Engineers. All rights reserved.

512779 Towards Sustainable Energy and Materials: Carbon Capture and Conversion using Novel Liquid-like Nanoscale Hybrid Particulate Systems

Abstract:

The atmospheric concentration of CO_2 has naturally fluctuated on the timescales of ice ages. Concerns, however, stem from the recent dramatic increase in CO_2 concentration, which coincides with global industrial development. This rise is mainly due to the high use of fossil fuels during power generation and chemical production. In order to meet the ever-increasing global energy demands while stabilizing the CO_2 level in the atmosphere, the development of carbon capture, utilization and storage technologies is one of the critical needs. In particular, there has been significant efforts to develop CO_2 capture solvents and some have shown very promising results. For example, amine-based aqueous solvents can effectively and selectively capture CO_2 from flue gas of coal-fired power plants. Unfortunately, the energy requirement for the current aqueous solvent systems is still considered to be too high. Thus, efforts have been focused on the development of second and third-generation CO_2 capture solvents which are often water-free. Nanoparticle Organic Hybrid Materials (NOHMs) are a new class of organic-inorganic hybrids that consist of a hard nanoparticle core functionalized with a molecular organic corona that possesses a high degree of chemical and physical tunability. NOHMs are liquid-like, non-volatile and stable over a very wide temperature range, which make them interesting materials for various energy and environmental applications. While their CO_2 capture efficiency and selectivity are great, like other anhydrous CO_2 capture solvents, NOHMs suffer from high viscosity. Thus, an innovative encapsulation system has been developed to create large gas-liquid interfaces for CO_2 capture using these viscous solvents and encapsulated solvents show greatly improved CO_2 capture rates. Furthermore, it has recently been discovered that NOHMs have interesting electrolyte properties which may allow the CO_2 capture to be pulled by the in-situ CO_2 conversion reactions. The development of these unique particulate systems will not only advance CO_2 capture materials design but also introduce unique particle technology research opportunities in various fields

Copyright © American Institute of Chemical Engineers. All rights reserved.

509490 An Industrial Perspective on the Future Needs in Solids Processing Research and Education

Abstract:

It is easy to get the impression that progress in the field of solids processing has been slow and plodding; however, if we examine the gains in particle technology over the last several decades, we should be proud of the collective accomplishments of the research community. Yet, many problems still persist â€“ old ones still needing a solution (ratholing and dense phase conveying, for example) and new ones that have surfaced as a result of recent technological advances and changes (additive manufacturing, advanced drug delivery, specialty materials and resource conservation to name just a few). This should delight the solids processing research communities in both academia and industry as there are many interesting, challenging and rewarding problems ahead of us. The key research problems across the 15-20 solids processing sub-disciplines will be examined from the viewpoint of their importance to particle technology, engineering science, industry, and more broadly, society as a whole.

Copyright © American Institute of Chemical Engineers. All rights reserved.

AUTHOR INDEX

AUTHOR INDEX